Mahtab Niknahad

Using Fine Grain Approaches for highly reliable Design of FPGA-based Systems in Space

Band 9
Steinbuch Series on Advances in Information Technology

Karlsruher Institut für Technologie (KIT)
Institut für Technik der Informationsverarbeitung

Using Fine Grain Approaches for highly reliable Design of FPGA-based Systems in Space

by
Mahtab Niknahad

Karlsruher Institut für Technologie (KIT)
Institut für Technik der Informationsverarbeitung

Zur Erlangung des akademischen Grades eines Doktor-Ingenieurs
von der Fakultät für Elektrotechnik und Informationstechnik des
Karlsruher Institut für Technologie (KIT) genehmigte Dissertation

Mahtab Niknahad aus Sabzevar, Iran

Tag der mündlichen Prüfung: 13. Juli 2012
Hauptreferent: Prof. Dr.-Ing. Jürgen Becker
Korreferent: Prof. Dr. Dr. h.c. mult. Manfred Glesner

Impressum

Karlsruher Institut für Technologie (KIT)
KIT Scientific Publishing
Straße am Forum 2
D-76131 Karlsruhe
www.ksp.kit.edu

KIT – Universität des Landes Baden-Württemberg und nationales
Forschungszentrum in der Helmholtz-Gemeinschaft

KIT Scientific Publishing 2013
Print on Demand

ISSN 2191-4737
ISBN 978-3-7315-0038-4

This thesis, for what it's worth, is dedicated to Olli for being an unlimited source of encouragement and love and to my grandma, who could not see the end, but whose memory has accompanied me giving me inspirations.

Abstract

Nowadays using SRAM based FPGAs in space missions is increasingly considered due to their flexibility and reprogrammability. A challenge is the devices sensitivity to radiation effects that increased with modern architectures due to smaller CMOS structures. This work proposes fault tolerance methodologies, that are based on a fine grain view to modern reconfigurable architectures. The focus is on SEU mitigation challenges in SRAM based FPGAs which can result in crucial situations. The two major approaches are fine grain TMR and Quadruple Force Decide Redundancy (QFDR). The classical concept of TMR and Quadded Redundancy on Logic Gates is transformed to an general approach that is based on FPGA structure primitives like LUTs. This work also shows, how these methodologies can be integrated into common CAD tools and tool flows in order to apply the techniques for different designs automatically. The results of this work show, that changing the granularity level of redundancy on SRAM based FPGAs brings benefits whenever high error rates are being considered. Eventually this work demonstrates how quadded logic even can be applied to future carbon nanotube based architectures beyond the CMOS roadmap.

Acknowledgments

I would not have been possible to write this doctoral thesis without the help and support of the kind people around me, to only some of whom it is possible to give particular mention here.

First and foremost, I would like to express my gratitude to my doctoral supervisor Prof. Juergen Becker. I am extremely grateful for his continuous support and guidance starting at the day, which he picked me up in the Frankfurt airport till the day of my PhD examination. He gave me a great opportunity to conduct my research in his group at the Institute of Information Processing Technologies (ITIV) and to explore my research interests with freedom and flexibility. His good advices, experiences and friendship, have been invaluable on both an academic and a personal level, for which I am extremely grateful.

I would like to thank my co-advisor Prof. Manfred Glesner at TU Darmstadt for accepting the role as co-advisor of my thesis and for providing me with important feedback and ideas for improvements. I am truly thankful for his ever-present support and encouragement in taking time and effort in reviewing my work.

Special thanks must be given to Prof. Uli Lemmer, as well as, Prof. Marc Weber, and Prof. Soeren Hohmann who kindly served as the members of my PhD oral defense committee. It has been truly a great privilege and honor to have them as mentors, and I received invaluable advice and constructive feedback for the work.

Further, I would also like to thank Dr. Michael Huebner for his never-ending source of ideas and inspiration. My work benefited a lot from discussions with Michael.

Furthermore I would also like to acknowledge the rule of Prof. Fernanda Lima Kastensmidt at Universidade Federal do Rio Grande do Sul, for the useful comments, remarks and engagement through the learning process of this thesis. Her contribution in stimulating suggestions and encouragement helped me a lot to coordinate my work.

My former room colleague Christoph Roth deserves an acknowledgment for his kindness and his enormous sense of humor. He has a special capability to make me laugh in any stressful situation.

My friends and colleagues at ITIV deserve my greatest thanks for the creation of a fantastic environment, which has been continuously exciting and full of joy. Further, I would like to thank the secretary group of institute, especially Mrs. Brigitta Hirzler, for their kindness and during the time I was in the institute.

I am thankful to my siblings, Mahnaz and Mahshid for their continuous support, encouragement and love through years.

My deepest gratitude goes to my parents, Maryam Hamzei and Mahmud Niknahad for all their sacrifice and care for my education and personal growth. I really appreciate them for their unconditional love and support, for giving me the freedom to choose and most importantly for being such great role models for me throughout my life. They instilled in me a thirst, a passion for education, learning, and for being the best that has stayed with me since childhood.

Finally, I wish to express a very special thanks to Oliver Sander for the last years immense support in both research related issues and personal. His wealth of ideas and breadth of knowledge as well as his love, selflessness and enthusiasm for science and humans have influenced me greatly. I learned a lot from his unique and hardworking character. Oliver, Thank you for being in this world.

Karlsruhe, May 2013

Mahtab Niknahad

Contents

1 Introduction

For a long time, human has fixed his eyes on the twinkling stars of the night skies. The mysterious darkness of the universe makes him think, what is out there? Is someone there? Flying out there has been an amazing wish of humankind.

In the twentieth century as spaceflight became a possibility, doors to this old wish have been opened. Many people are now considering the future of space researches in order to find some answers for their questions.

This contribution, focuses on the challenge of radiation effects in space for Field Programmable Gate Arrays (FPGAs). FPGAs are reconfigurable architectures which are suitable in space application.

1.1 Problem Definition

The motivation that led to this work is the increasing consideration of FPGAs for space environments [101] [24]. The decreasing cost and development time which is needed to implement FPGAs as well as their flexibility in fulfilling modification requirements and remote programmability are the key factors toward their success in space and avionic applications.

Another factor that attracts even more attention to FPGAs in space is their simplicity. When compared to Application Specific Integrated Circuits (ASICs), tooling for FPGAs is less complex and also cheaper as the manufacturing steps can be omitted. Especially in space environments which high volume is not an issue, ASIC design may be very expensive.

Even for the applications which are run in space, FPGAs can play a stronger role in comparison to ASICs. For example, in processes, like geography or weather forecasting, image processing is an important processing step which can be done in orbit. Due to their flexibility, hardware implemented image processing algorithms can be exchanged on demand from the ground station. Due to this fact, using FPGAs for these applications makes sense [72].

Therefore, the first part of motivation involves using FPGAs with their reconfigurability features in space applications and in the satellites which are in orbit or in space. However FPGAs encounter challenges in space. Although FPGAs are

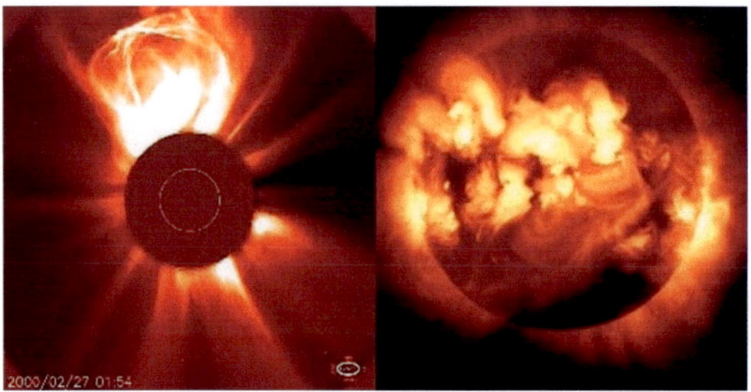

Figure 1.1: (left) Coronal mass ejection and (right) multiple solar flares make solar winds.[108]

highly flexible, they are sensitive to radiation effects which exist in space [78]. Overcoming this issue is the second motivational part of this work.

In our solar system, the major source of radiation effects is the sun. It emits particles continuously. It consists of gas and due to it is less cohesive [108]. However, its rotation is not like a solid ball such as earth. It is a process of rotating the gas mass which may result in coronal mass ejection or in solar flares(Figure 1.1).

Both of these events produce massive amounts of electrons, protons, and photons which have an impact on every thing they meet. They can generate solar winds. These are the source of radiations which can effect integrated circuits temporarily or permanently.

The effect of radiations on the chips can even be stronger these years as the CMOS scaling has pushed the feature size smaller and smaller in nano scales. Therefore, irradiation has a bigger effect on the nano scale chips [112] [43].

The goal of this work is to develop a methodology that allows for making designs reliable for space applications automatically. Although several mitigation techniques exist that target FPGAs reliability, there are some scenarios in modern designs, where upsets can propagate in design and make it fails in critical space applications. Accordingly, new methodologies for mitigation techniques are needed to cope with the radiation effects in nanoscale designs.

The result of these investigations led to the fine grain fault tolerance approach for reliable design. This includes both using traditional approaches like redundancy techniques but also novel ones where the characteristics of FPGAs including homogeneity is exploited.

With respect to Moore's Law[110], which mentions the end of road map for CMOS technology, nano architectures and their features are currently studied.

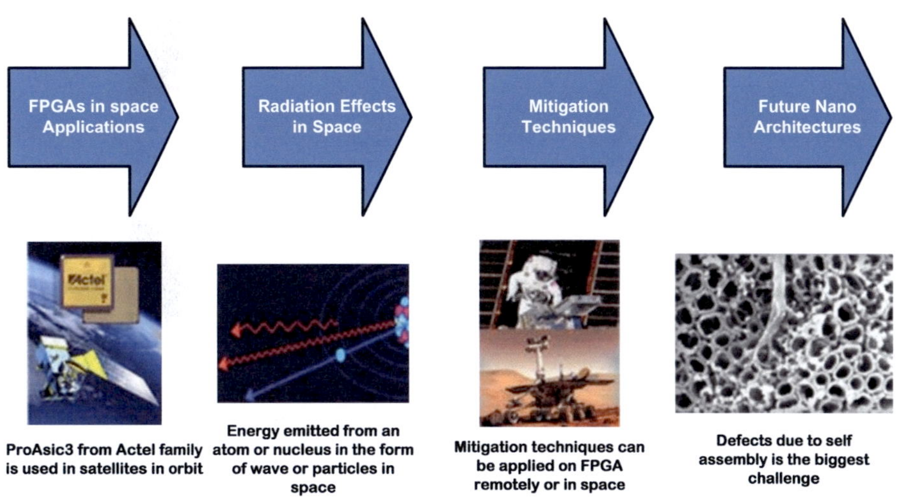

Figure 1.2: Main steps towards using FPGAs in Space

They are the strongest alternative to CMOS technologies in future [12][29][39]. These architectures are unreliable by nature. Their self assembly process of construction results in a lot of defects [29]. In this work, nanoPLAs as an alternative to FPGAs are studied in specific. As result fault tolerance methodologies for FPGAs are modified, to be used for defect tolerance goals in nanoPLAs.

1.2 Objectives

As mentioned, successful usage of FPGAs in space depends on reliability and hence the mitigation methodologies which are used in order to mask or correct any kind of failure caused by a radiation effect. The main objective of the work presented here is reliability improvement for FPGAs in space. Therefore, the path toward FPGAs usage in space includes three major steps that are depicted in Figure 1.2.

Future nano architectures cope with some challenges in order to be a suitable alternative architecture to FPGAs. This work targets the defective nature of nanoPLAs due to self assembly and modifies defect and fault tolerance techniques for the sake of reliable design on nanoPLAs.

1.3 Outline of this Work

The thesis is organized as follows:

Chapter 2 introduces FPGAs and their architecture including their history and evolution. It describes different kind of FPGAs and their features. It explains the advantages of FPGAs to ASICs in different applications including space applications.

Chapter 3 introduces radiation effects in space in detail. The resource of radiations as well as the types of effects which can be harmful for semiconductors especially FPGAs are discussed there. It presents the major effect categories, which upset can harm FPGAs, and discusses its ability for different kind of FPGAs.

Chapter 4 presents the state of the art of the current mitigation techniques. These techniques basically are divided into two different categories: reconfiguration-based and redundancy-based techniques. In this chapter, these two categories are discussed in detail. The methodology of this work is based on redundancy-based technique. This technique is studied in detail and its advantages and disadvantages which lead to new approach are described.

Chapter 5 targets the new redundancy-based technique which is called QFDR. It introduces the history of the methodology for basic logic gates and the ideas which modifies it for modern FPGA technology. However, its application on FPGAs and the manner of realizing it on FPGAs is the topic of Chapter 6. Chapter 6 also includes the results on area, fault tolerance and power usage of new methodology on target FPGA.

Chapter 7 focuses on new architectures at the end of road map. While reasoning nanoPLAs as a good alternative to FPGAs in future architectures, the modified methodologies in chapter 5 for defect tolerance reasons on the designs in nanoPLAs are applied.

Finally, Chapter 8 concludes the thesis and presents a few suggestions for future work.

2 Modern FPGAs for Space Applications

A Field Programmable Gate Array (FPGA) is a semiconductor logic device with programmability. Programmability in FPGAs makes it possible to realize a logic function after the manufacturing process. One can write the desired logic in a high-level hardware language like VHDL or Verilog and finally program it on FPGA after an automated process. FPGAs can be configured one time or multiply. The SRAM-based ones can be reconfigured practically unlimited times. In this chapter, a brief history of the FPGAs and the state of the art FPGAs is summarized. It will provide the background for further discussions on their usage in space applications.

2.1 Field Programmable Gate Array

FPGAs, as illustrated in Figure 2.2, consist of an array of reprogrammable logic gates as well as distributed memory blocks, that all are connected through a hierarchy of configurable interconnects.

Modern FPGAs also provide embedded IP cores, such as memories, DSP blocks and processors in order to achieve the implementation of System On a Chip

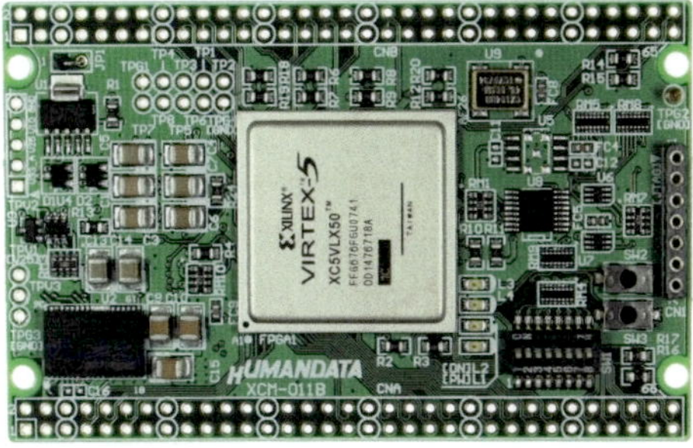

Figure 2.1: a Xilinx FPGA chip

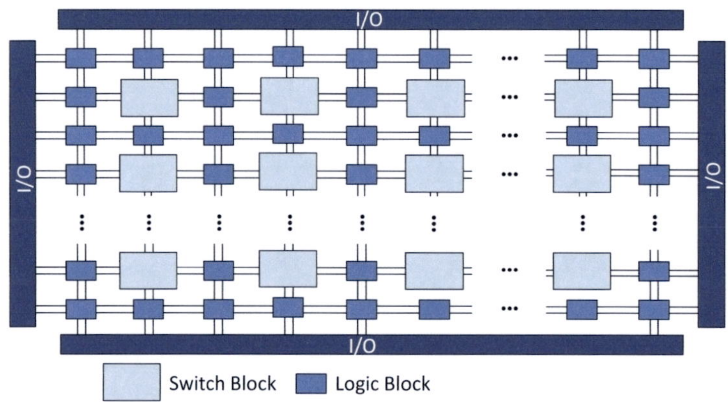

Figure 2.2: Logic Structure of an FPGA

(SoC) designs. A picture of a Xilinx FPGA chip is depicted in Figure 2.2. Its simplified architecture is reflected in Figure 2.2. The chip is surrounded by input/output blocks, which are briefly called I/Os. I/Os connect the FPGA to other devices on the printed circuit board (PCB). The very first programming technology used small fuses. As the FPGA architecture evolved and its complexity increased, programming technologies has become more sophisticated.

The history of FPGA returns back to the history of developing the integrated circuits in the middle of 20th century. The need for getting designs quickly done, was a big motivation to evolute Field Programmable devices. The first senses of programmability were the series of Read Only Memory (ROM). The basic ROM is a single one time programmable logic Array. Other variations of the ROM includes more flexibilities in programming. PROM is a version of ROM which is one time programmable. The difference between it and the basic ROM is the way of programmability. In the basic ROM, programming is done by masking in IC manufacturing process. PROM allows one time programming by user[88].

EPROM is the erasable PROM which can be completely erased and reprogrammed. However, writing on EPROM is generally slower than reading from it. This is while not considering the time for reading the entire EPROM. EPROMs (including ROMS and PROMs) use N address input to implement a function. They are able to implement any logic function with N-input. However, by growing the N, the area usage increases exponentially and this is an issue for these devices [88].

To cope with area overhead, the logic function can be restructured to levels of AND/OR gates. In this way, which is significantly more area efficient, wired OR, wired AND planes, and inverters are used to build logic terms (Figure 2.3(a)).

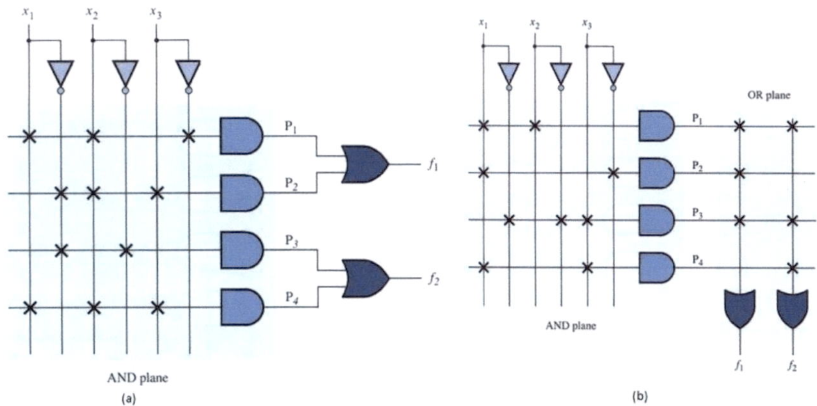

Figure 2.3: PLA Structure[1]

2.2 Evolution of FPGAs

Programmable array Logic(PAL) devices use the programmable AND planes which are followed by the OR planes. This is shown in Figure 2.3(b). This provide flexibility to program any combinational or even a kind of sequential logics by programming D-type flipflops on planes.

Unlike the simple logic functions, in case of data paths and multi level logic circuits, interconnections can be a problem for PALs. Device inputs and intermediate combinational sums are fed into the array via a programmable interconnect. It must be a typical full crossbar to program every level. This adds a significant interconnection (area) cost to the circuit in case of data paths or multilevel logics [77].

To achieve flexibility and area efficiency as well, Static Memory based FPGA were proposed. This architecture uses bit streams to configure either logic or interconnections. Logic cells are the elementary part of FPGAs. They are used to implement logics and storage elements. Additionally, there are some inter-cell connections which are flexible and can be changed by configuring the bit stream. This enables the implementation of different kind of complex circuits including complex SoCs.

However, there is always a trade off between the area usage and the flexibility. Static memory offers flexibility in programming and at the same time increases area usage per programmable switches compared to ROMs. As also [77] mentioned, this was a key issue in delaying the introduction of commercial SRAM-based devices till the cost per transistor was sufficiently lowered at the mid of 1980's.

Xilinx was the first company which introduced the modern era FPGAs [77]. The structure of modern FPGAs includes an array of Configurable Logic Blocks, that initially go back to the first FPGAs. These contained 64 logic blocks and 58 inputs and outputs [77][74].

After that, FPGAs have grown enormously in complexity with every generation. Modern Xilinx FPGAs include more than millions of logic blocks, huge number of inputs and outputs in addition to a large number of specified blocks which makes SOC possible. This significant improvement in the capabilities of FPGAs includes the massive architectural changes.

2.3 State of the Art FPGA technologies

Different programming technologies have been introduced for FPGAs. These technologies primarily rely on controlling the programmable switches which are used in FPGAs. Different approaches historically include EPROM [47], EEPROM [51], flash [57], static RAM [44], and antifuses [25]. Among them, only the flash, static RAM and antifuse approaches are still used in modern technologies [77]. Due to its features, this work focuses primarily on SRAM-based FPGAs. This section considers current programming technologies in oder to provide a more understanding of the advantages and disadvantages of SRAM-based programming.

2.3.1 SRAM-based FPGAs

SRAM programming technology uses static memory cells or SRAM cells for programming. This technology has been widely used in devices which are made by Xilinx[9], Lattice [6], and Altera[4]. In SRAM-based devices, static memory cells, such as the one shown in Figure 2.4, provide configurability for FPGAs. SRAM cells are used in interconnection or implementing logic functions. In interconnects, SRAM cells are used to select lines to multiplexers in order to connect logic components in an appropriate way. In order to implement logic functions, SRAM-based FPGAs use Look Up Tables(LUTs). Figure 2.6(a) and Figure 2.6(b) illustrate the structure of SRAM cells in MUX.

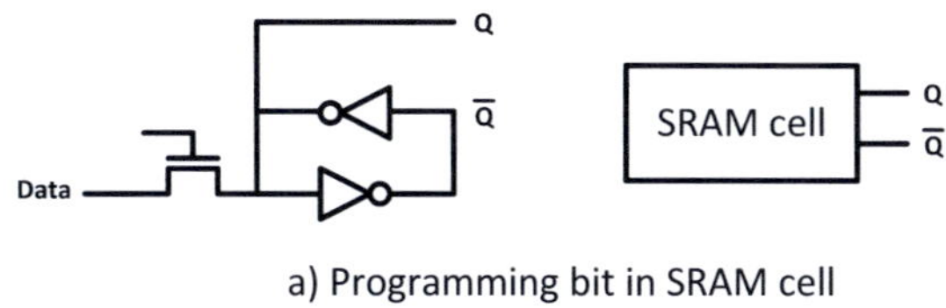

Figure 2.4: Program Bit in SRAM Cell

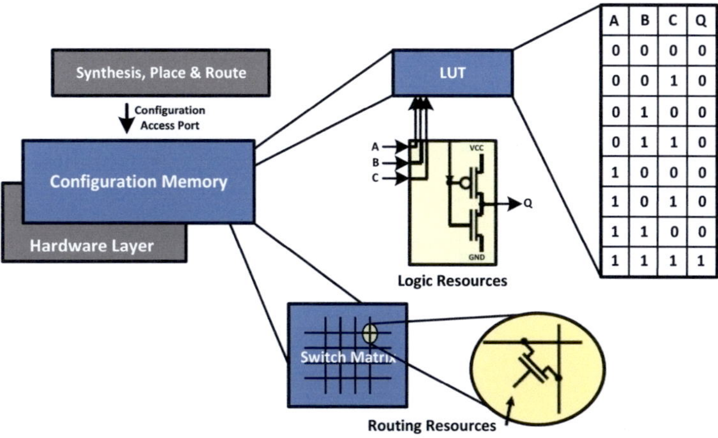

A	B	C	Q
0	0	0	0
0	0	1	0
0	1	0	0
0	1	1	0
1	0	0	0
1	0	1	0
1	1	0	0
1	1	1	1

Figure 2.5: Structure of Configuration in SRAM-Based FPGAs

SRAM-based FPGAs are widely used in different applications due to their repro-grammability. SRAM bits can be configured by user defined configurations on power up. This can be done an indefinite number of times. Unlike other pro-gramming technologies, SRAM-based cells use the standard CMOS processing technology. As a result, the latest available CMOS technology can also be applied to SRAM-based FPGAs and, therefore, benefit from the increased integration, the higher speeds and the lower dynamic power consumption.

The configuration memory controls the logic implemented by the FPGA device as well as the interconnect between logic functions in routing. Configuration is done and defines arrangement of pre-existing logic, via programmable switches, which includes the function of system in logic clusters, the connectivity as the routing and placement of the design on FPGA. Configuration on FPGA can be done either Antifuse, which is one time Programmable, Flash-based or SRAM-based, which are re-programmable. Figure 2.5 shows structure of configuration in FPGAs.

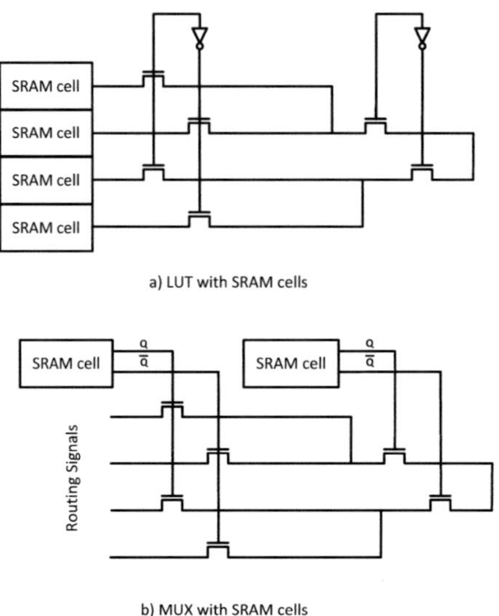

Figure 2.6: LUT and MUX by SRAM cells

However, there are some characteristics which are a challenge for SRAM-based technologies. [77] mentions some of obstacles of SRAM-based FPGAs. One of the most important drawbacks is the volatility. When an SRAM-based device is powered down, it needs the use of external devices to store the configuration permanently. Recently, flash memories in some devices are used to load SRAMs upon power up. This means that there are a few devices that have solved this problem, but the need for flash memory make them inefficient [77].

The other issue is the size of SRAM cells and interconnection signals respectively. SRAM cells require 5-6 transistors and the programming elements which are used for interconnecting signals need at least one transistor.

Configuration informations are loaded into the device at power up. Since the configuration information is not encrypted (or secured) it can easily be used by the other systems.

Finally, the use of pass transistors in implementing multiplexers can make some problems due to the electrical properties of pass transistors. They have a significant on-resistance and an appreciable capacitive load. Consequently, smaller device geometries make this issue more critical.

2.3.2 Antifuse-based FPGAs

Programmable switches in FPGAs can be implemented in Antifusebased manner. As previously mentioned Antifusebased FPGAs are one time programmable. This is depicted in Figure 2.7. There is a highly resistive amorphous semiconductive layer in Antifused-based switches which is irreversibly switchable to a low resistive state. The programming method in antifuse is done by growing a link to make a connection instead of passing current through a metal connection and breaking it.

Dielectric antifuses are composed of an oxide nitride which is positioned between N+ diffusion and polysilicon. [77][58]. The application of high voltage breaks down the dielectric and form a conductive link. This link has a resistance between 100 and 600 ohms [77][34][56]. This dielectric approach has been largely replaced by metal-to-metal-based antifuses. These antifuses are formed by sandwiching an insulating material such as amorphous silicon [54] or silicon oxide [36] between two metal layers. Again, a high voltage breaks down the antifuse and causes the fuse to conduct.

Consequently, antifuse technology is suitable to be used in creating programmable interconnects. However, it require large programming transistors on the device. Because this FPGAs can not be reprogrammed, design changes are not possible and, configuration retains after power off hence making their configuration immune against single Event Effects [69][109][67].

One of the most significant advantages of anti fuse programming technology is that it is not area hungry. One of the benefits of metal-to-metal antifuse is that the on resistance can be between 20 and 100 ohms[111]. As a result, with metal to metal antifused no silicon area is required for connection. This decreases the area overhead of programmability. The metal-to-metal approach used recently in FPGAs is from Actel[2] and Quick Logic[8].

Although the metal-to-metal approach decreases the area usage in antifused programming, the need for large programming transistors, almost offsets it. Large current needs to program antifuse and large transistor must be used to be able to supply it. Since clever programming architectures are used for fuses, the total area can be significantly decreased. Another advantage of antifused-based technologies is lower residence and parasitic capacitances in comparison to other programming technologies like SRAM-baseds. Therefore, more switches can be made in antifused devices.

Because of the one time programmability, the volatility is not a problem for antifused technologies. Device operates immediately upon power up, and it does not require an additional memory to store programming information or configuration time. This significantly decreases the overhead and especially the costs.

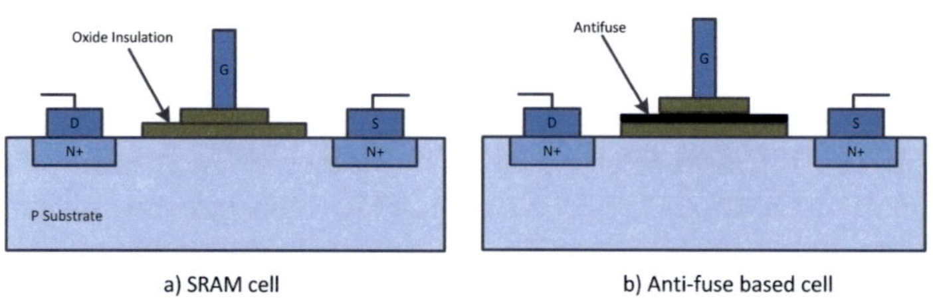

Figure 2.7: Antifuse vs. SRAM-based FPGAs

With regard to security, antifused-based devices show an improvement. Because they are one time programmable, transmitting the bitstream to the FPGA is done only one time. This improves the security of the design on the FPGA while the programming informations are not available for further uses. Current devices use a security mode, wherein they disable any access to the programming interface as soon as the device is programmed[77].

Almost all advantages of SRAM-based technologies are included in disadvantages or inabilities of antifused-based ones. In particular, the antifused-based FPGAs need nonstandard CMOS processes. It means that, they are unable to use the latest technologies which are available for CMOS. Therefore, they need a particular manufacturing process in comparison to SRAM-based FPGAs.

Furthermore, the mechanism of programming includes significant changes of the material in the fuses. This means scaling in new IC fabrication processes is a challenge and costable.

One of the biggest limitations in antifused-based technologies, which is also the main property, is their one time programmability. This makes them inflexible for any design change. For example, they are especially unsuitable for applications where configuration changes are required. Unlike alternative technologies, in-system programming is not possible with these devices. Instead, special programmers must be used to program a device before it is mounted on a final product.

Finally, the one-time programmability of antifuses makes it impossible for manufacturing tests to detect all possible faults. Some faults will only be uncovered after programming and, therefore, the yield after programming will be less than the 100% yield of SRAM or floating-gate devices. Some current devices are only expected to be programmed successfully with 90% confidence[77].

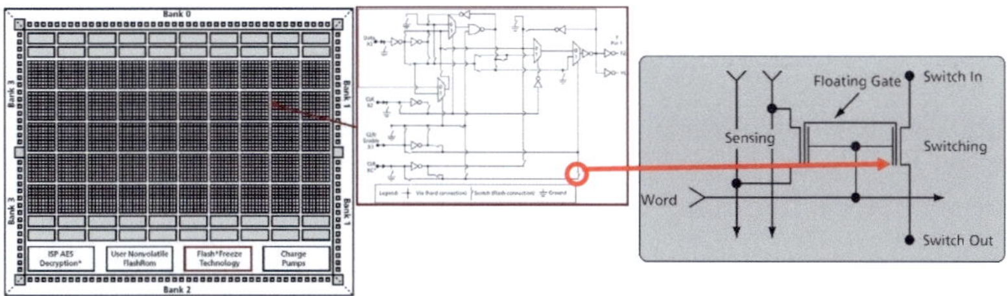

Figure 2.8: A ProASIC3 from ACTEL family which is a flash-based FPGA. Switches use floating gates [3].

2.3.3 Flash-based FPGAs

One alternative to SRAM-based technology, which eliminates some of their obstacles, is the use of floating gate programming technologies that inject charge onto a gate that floats above the transistor. This approach is used in flash or EEPROM memory cells. These devices, as described, do not lose information when they are powered down.

In spite of this ability, EEPROM memory cells commonly have been used to implement wired-AND functions in PLDs instead of implementing switches in FPGAs. [104].

However, as also [77] mentions, except for very low-capacity devices [5], such approaches are no longer commonly used. The most important reseal is the static power dissipation. In addition to it, in modern IC fabrication processes, the use of floating gate cells directly as switches has become possible.

Flash memory cells, in particular, are now used because of their improvements in area efficiency. Figure 2.8 illustrates Actel's ProAsic device which uses flash based technology for programming. In this device, larger transistors are used in programming switches and small transistors are used to program the floating gates. When the device is powered off, the injection charge will remain. When erasing the device, the switching transistors(larger one) are used [3].

Flash based programming technology offers several unique advantages. The most important one is non-volatility. In addition, like antifused-based technologies, these devices are free from any kind of external memories to store the programming information. This increases security, as well.

As mentioned, SRAM-based devices need to wait at power up for loading the configuration data. In spite of it, the flash-based devices function immediately after power up. In size issue, SRAM-baseds need six transistors to implement the programming storages. However, programming circuitry, which is needed to program the cell (for example high/low voltage buffers) is also included in area overhead. Flash-based technologies have less than SRAM-baseds area overhead [77].

In comparison to antifuses, an alternative non-volatile programming technology, flash-based FPGAs are reconfigurable and are able to be programmed more than one time without being removed from a circuit board.

In order to control the switching transistors with floating gates, the source-drain voltage must remain sufficiently low to prevent charge injection into the floating gate [85]. This adds some design complexities to the system. However, this issue may be solved in the near future by newer processes which require lower voltage levels.

Flash-based devices cannot be reprogrammed an infinite number of times. This seems to be a significant disadvantage. Charge buildup in the oxide eventually prevents a flash-based device from being properly erased and programmed. Current devices such as the Actel ProASIC3 are only rated for 500 programming cycles [77]. However for most uses of FPGAs, this programming count is sufficient. In many cases, (but not our case in space applications) FPGAs are programmed for only one use.

Another significant disadvantage of flash devices, like antifused-based ones, is their need for a non-standard CMOS process. This was described in detail for antifused-baseds and can be applied also to flash-based technologies.

Also, like the static memory-based technology, because of using the transistor based switches, this programming technology suffers from high resistance and capacitance.

One trend, which is recently taken into account, is the use of flash storage in combination with SRAM programming technology [79]. In these devices from Altera, Xilinx and Lattice, on-chip flash memory is used to provide non-volatile storage while SRAM cells are still used to control the programmable elements in the design. This addresses the problems associated with the volatility of pure-SRAM approaches, such as the cost of additional storage devices or the possibility of configuration data interception, while maintaining the infinite reconfigurability of SRAM-based devices.

It is important to recognize that, since the programming technology is still based on SRAM cells, the devices are no different from pure-SRAM-based devices from an FPGA architecture standpoint. However, the incorporation of flash memory generally means that the processing technology will not be as advanced as pure-SRAM devices. Additionally, the devices incur more area overhead than pure-SRAM devices since both flash and SRAM bits are required for every programmable element. Figure 2.8 shows a Flash-Based FPGA(ProASIC3) [3] from ACTEL family in simplified and detail form.

Table 2.1: Summary of State of the Art Technologies

Technology	Reprogrammable	Volatile	Technology
Antifuse-based	no	no	non Standard CMOS Tech
SRAM-based	yes (in circuit)	yes	CMOS Tech
Flash-based	yes but finite	no	non Standard CMOS Tech

2.4 Field of Applications

In this section a non exhaustive list of fields where the use of FPGAs can be a great interest is presented. Because FPGAs are still in growing edge, several different fields are likely to be developed in the near future, which were not conceivable in the past [26].

In comparison to the standard cell ASICs, FPGAs are more power, area, and performance hungry. [77] mentions 20 to 35 times more area usage and 3 to 4 time slower in speed performance is compare to ASICs. These disadvantages arise largely from FPGA's programmable routing fabric which trades area, speed, and power in return for instant fabrication.

Despite of these disadvantages, the trade off is getting flexibility. They are naturally more flexible and can be used in a wide area of applications due to their flexibility. FPGAs present a compelling alternative for digital system implementation as they have a low volume cost for small to mid volumes. Producing a design on ASIC consists of several complex processes which means times and money. FPGAs provide the only economical access to the scalability and performance provided by Moore's law.

Rapid Prototyping is certainly one of the most important fields of application of FPGAs. The development of ASIC that can be seen as physical implementation is a process consisting of several steps, from the specification down to the layout of the chip and final production. FPGAs are useful here, because they can be used several times to implement different versions of the final product checking it to be in an error free state.

In addition, for complex applications, like security applications, FPGAs are useful in developing a fast system. FPGAs also provide a good fundamental for the realization of adaptive systems, because they allow the system to quickly react to changes by adopting the optimal behavior for a given runtime scenario.

In-system customization can be used to upgrade systems that are deployed into non-accessible or very difficult access locations. FPGAs are the most suitable devices in the case, because they can be modified remotely. This in addition to the above features makes FPGAs specially good for space kind applications.

Figure 2.9: FPGAs are expected to be widely used in Space Applications.

2.4.1 FPGA technologies in Space

FPGA devices have been used in space applications for more than a decade. The capacity and performance of FPGAs, in addition to their reprogramming flexibility, increases use of them for space applications steadily. Figure 2.9 shows a satellite in orbit which is covered by FPGAs from Xilinx.

In FPGAs, the capacity has grown up from tens of thousands to millions logic gates. The usage of FPGAs has transfered from simple glue logic to complete subsystem platforms that combine several real time system functions on a single chip, even including microprocessors and memories [26].

Not only in space, but also, in any other type of critical applications, FPGAs usage is steadily increasing and are replacing ASICs on a regular basis.

The potential of using reprogrammable FPGAs in space has been presented in [47] and is repeated hereafter. FPGAs Reconfigurable computing technology is still a new field of study for space applications.

Space environment is different from terrestrial systems. In terrestrial systems, radiation can cause bit flips in memory elements and as a result ionization failure in semiconductors. This kind of hardware faults cannot be debugged and repaired. This means that terrestrial systems need a high-reliability manufacture in order to avoid such cases.

FPGAs on the other hand, eliminate this need by reconfigurability. The use of run-time reconfiguration in space will allow to change on-board hardware by replacing faulty/outdated designs in a space mission. They can be used in tasks that are formerly handled at the board levels by separated, dedicated parts. FPGAs are sufficient to implement a system on chip. This eliminates the need for extra components phase lock loops, voltage translation buffers, and memory when on-chip memory is sufficient. This high level of integration and flexibility extremely increases the cost and other system requirements in critical space applications [55].

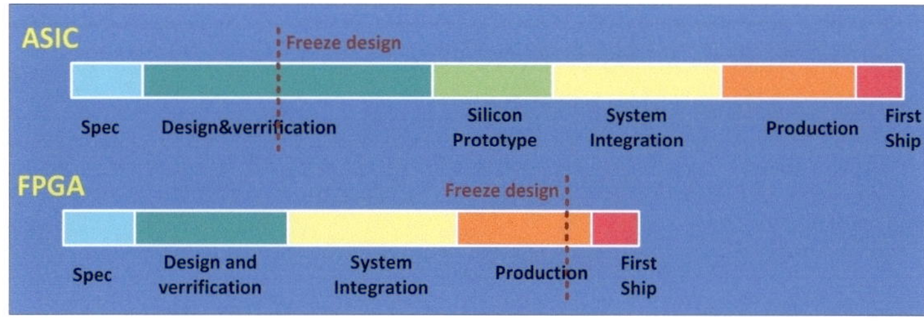

Figure 2.10: Design Steps in FPGAs and ASIC. FPGAs allow late design changes.

2.4.2 ASICs vs. FPGAs

ASICs are customized integrated circuits designed to perform a dedicated function. They can be designed using three main techniques: Standard Cells, Gate Arrays, and Full Custom. Standard Cell employs third party tools and utilizes a library of prefabricated building blocks (the so called standard cells) to design the integrated circuit. Once the physical placement of all standard cells is done, the designer must perform the circuit routing therefore, electrically connecting all the standard cells. At the end, lithographic layers are produced, which are then used to fabricate the chip die.

Cost of ASIC design and developing is high [76]. Normally Space Applications are single usage products. ASICs are targeting big volumes to make it lower cost. By Unique application staffs this big volume in production is not possible. Therefore the cost is even more in general. On the other hand by physical damages, they must be replaced with new ones. This means a big overhead on space missions.

In contrast, Gate Array design does not employ individual cells. Instead of that, several pre-designed lithographic layers are employed, each of them consisting of transistors, gates and other devices. In this technique, all the appropriate elements must be connected together to obtain the desired circuit functionality. Alternatively, the Full Custom approach requires all elements of the entire lithographic layers to be designed.

In case of FPGAs it is easier and more flexible to reprogram them in case of failures in the system. When a damage occurred, the reconfiguration is done by avoiding the damaged spaces on board or changing the reconfiguration in a way that the system works properly. Also, usually in space application there is an extremely vital need of data processing. By using FPGAs which are faster in data processing, the cost of data transferring into and from space is saved. Also, the capacity and performance of FPGAs suitable for space flight have been increasing steadily for more than a decade. As a result FPGAs are a technology that will be used in space applications in the coming years. Therefore, FPGA technologies continue to advance by NASA flight projects as either low-cost or schedule-effective alternatives to ASICs[107].

3 Space Applications and Soft Errors in FPGAs

3.1 FPGAs usage in Space Applications

3.1.1 Radiation Challenge in Space

Radiation effects on semiconductor devices have always been a challenge since upsets were first experienced and detected in space applications some decades ago. Therefore, the interest on studying fault tolerance techniques in order to protect integrated circuits in space environment has increased.

Integrated circuits have a variety of usage in space applications. They can be a part of digital components, which are used in satellites or spacecraft system. Digital components are potentially sensitive to radiation and must be protected and kept operational in space operations. Single event effects, which are known as soft error, are the main concern in space applications. Soft Errors induced by radiation, are becoming a more significant challenge since the scales in modern integrated circuits decreased to nanometers.

3.1.2 Mechanism of Radiation Effects

In order to become familiar with different radiation effects, some facts about the solar system are needed. The sun in the center of the solar system is a star which is a hot ball of glowing gases [10].

At the core, its temperature is about 15 milion degrees Celsius, which is sufficient for sustaining nuclear fusions. Energy from the core takes about 179,000 years to get from the core to the surface. The temperature drops below 2 milion degrees Celsius in the surface [10]. It is still high enough. The particles gain sufficient energy and are able to escape the gravity force of the sun. They travel to the outer space and reach the Earth's Magnetosphere. Solar winds are composed of protons, electrons, alphas ions, and heavy ions.

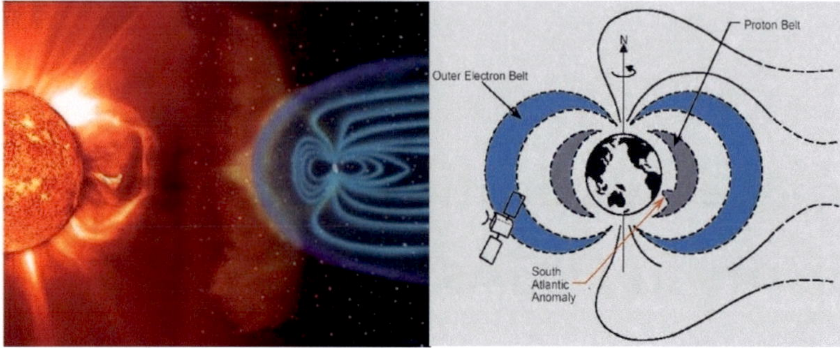

Figure 3.1: Van Allen Belt

The Earth's magnetic field or geomagnetic field extends from Earth's inner core to the point that meets the solar wind. The Earth is largely protected from the solar wind by its magnetic field which is always in interaction with the solar wind. This region is not uniformly distributed. On the side, which is facing to the Sun, it is compressed, while on the other side it is elongated as is shown in Figure 3.1.

The Earth's magnetic field is like a belt around the inner core. It was firstly predicted by James Van Allen, American astrophysicist and is called Van Allen Belt. The Van Allen Belt, as shown in Figure 3.1, consists of two regions, which trap particles in advance. The inner belt contains protons while it is separated by a region of reduced particle flux from the outer electron belt [46].

Outside the Earth's atmosphere, objects are always threatened by energetic particles in the form of solar winds. On the ground, electronic systems are protected from most of the particles by the atmosphere covered by the Van Allen belt [17]. However, particles can provoke serious damage to electronic crafts, which are traversing orbit and outside of the belt.

3.2 The SEE Problem

Radiation effects on on-board electronics are roughly divided into two categories: total ionizing dose (TID) and single event effects (SEEs).

Total ionizing dose (TID) is a long term damage effect on the device. It is cumulative and starts effecting the device when it is exposed to ionizing radiation. TIDs define the total sum of radiation hitting the target component in duration. It results in threshold shifts, increased device leakages, timing changes, and decreased functionality.

Gamma Rays primarily contribute to TID accumulated over months and years in orbit. Accelerated TID testing can be done by using a Cobalt source. The cumulative radiation effects from Cobalt are the most similar one to those experienced in space [61].

Device shielding can reduce the TID, but several factors must be studied and considered. Shielding is usually used to protect the device against TIDs. But parameters like shield geometry, material composition, and shield analyzing techniques must be taken into account in predicting shield effectiveness. For example, if aluminum shielding is used, electrons are effectively depressed even in very high energies, but the same aluminum shielding can be ineffective for the high energy protons.

Single event effects are radiation effects which occur when a single incident ionizing particle deposits energy and this energy is enough to affect a device [10]. Unlike TID degradation, SEEs are not a long term process and due to this the rates are not evaluated in terms of a time or dose until failure.

SEEs can occur in many forms. They can occur either on ground level or in space. The interaction of cosmic rays with Nitrogen and Oxygen in atmosphere can produce neutrons and other kind of particles called secondary particles. These particles, as well as the ground level generated particles may change the voltage or current in part of a circuit while passing through it.

Depending on the amount of deposited energy different effects occur. The success of any device in a critical space mission depends on the functional impact of SEE in addition to the probability of the occurrence based on the system structure and mission orbit. These issues are further used to equip a system [10].

3.2.1 Types of Single Event Effects

Single Event Upsets (SEUs) are soft errors, and non-destructive. They normally appear as transient pulses in logic or support circuitry, or as bit flips in memory cells or registers.

Several types of hard errors, potentially destructive, can appear: Single Event Transient (SET) which are transient upsets in logic and potential SEUs, Single Event Latchup (SEL), single event burnout (SEB), and single event gate rupture (SEGR) result in a high operating current, above device specifications, and must be cleared by a power reset.

Other hard errors include Burnout of power MOSFETS, Gate Rupture, frozen bits, and noise in Charge Coupled Devices (CCDs) [99][10].

SEUs and SETs are the main concern on SRAM-based FPGAs [28] [99] [81]. In the following, they are briefly described:

1. **Single Event Upsets**: Single Event Upsets (SEUs) are a kind of soft errors caused by the transient signal induced by a single energy particle strike. When a particle causes a charge disturbance and it is large enough to reverse or flip the data state of a memory cell, register, latch, or flipflop, an SEU occurs [106].

2. **Single Event Transients**: A Single Event Transient (SET) is a transient pulse in the logic path of a device. Similar to SEU, it is induced by a charge deposition of a single ionizing particle. An SET can potentially cause an SEU as it can be propagated along the logical path. When a flipflop in the path latches it, an SEU occurs. This makes them very important in reliability evaluation approaches.

A single particle strike may also be an SEU which affects multiple memory cells or results in multiple bit flipping. In this case it is called MBU. This work focuses on SEUs, SETs, and MBUs on SRAM-based FPGAs.

3.3 SEE characterization

In order to express the SEU characterization some parameter definitions are needed. In the following these definitions are gathered:

1. **Linear Energy Transfer (LET)** is a measure of the energy transferred to the device per unit length as an ionizing particle travels through a material. In general LET is defined as follow:

$$LET = \frac{1}{\rho} \frac{dE}{dx} \tag{3.1}$$

   Which means, the Energy per dx unit of a material with density of ρ. The unit of measurement arise from a combination of the energy lost by the particle to the material per unit path length MeV/cm and the density of material which is $\rho = mg/cm^2$. LET unit is defined as division of these units. Therefore, the common unit is $MeV\frac{cm^2}{mg}$.

2. **LET threshold** (LET_{TH}) is the minimum LET to cause an effect. The JEDEC recommended definition is the first effect when the particle influence = $10^7 ion/cm^2$ [45].

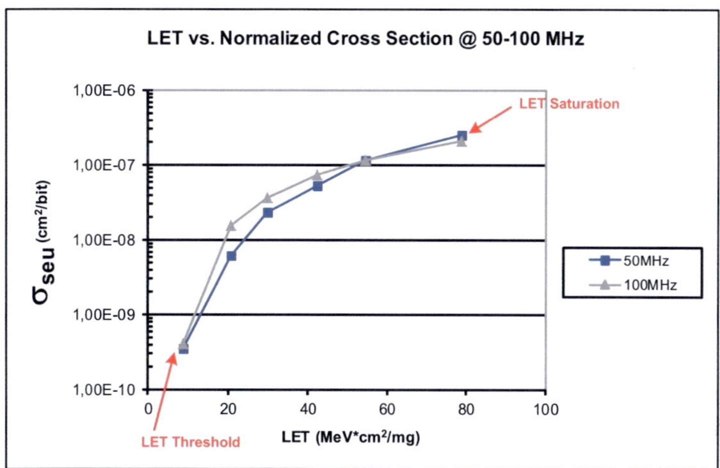

Figure 3.2: LET vs. the cross section of SEU based on the Weibull formula measured in two frequency domains. LET Threshold, is the minimum amount of deposited energy likely to induce an upset. Under 20 MeV, components are generally not SEU sensitive to incident protons. Beyond 300 MeV, the cross-sections generally reach a saturation value. [45]

3. **Cross section (σ)** characterizes the number of upsets, which occur based on the number of particles the device is exposed to. It is the device SEE response to ionizing radiation. For an experimental test for a specific LET, $\sigma = \#errors/(ion_in fluence)$. The units for cross section are cm^2 per device or per bit.

3.3.1 SEUs vs LETs

The influence of radiation is measured by cross section of every SEU on the device. Several analysis and experiments have been done in order to characterize it. The results show that, the σ_{SEU}, has an exponential relation with LET. [125] has done some experiments on Proton strikes on Xilinx FPGAs, and the bit cross section versus the LET is as Figure 3.3.

[53], [84] and [97] used Weibull formula based on the Weibull Distribution [124] and determined the length from heavy ion upset cross section versus LET in general. A Weibull distribution is a mathematical description of the failure behavior in a population of identical components. This distribution can be used as a reasonable model in order to describe device failures due to single events. The

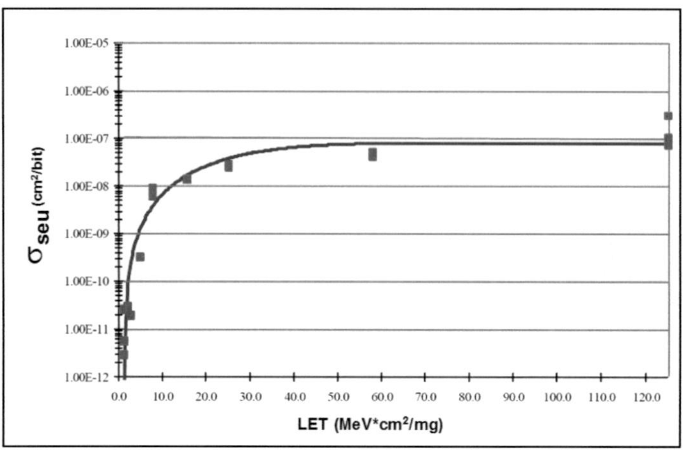

Figure 3.3: Static heavy ion bit upset cross-section vs. LET for Xilinx Virtex XQVR300 [49]

Weibull distribution defines the cross section versus LET curve. It is a function with four parameters (Equation 3.2).

$$\sigma\left(L\right) = \sigma_{sat}\left[1 - exp\left(-\left(L - L_0\right)/W\right)^{S}\right] \tag{3.2}$$

In Equation 3.2, σ_{sat} is the Asymptotic Saturation Cross section, L_0 is the absolute LET Threshold, W is the statistical width, and s is the statistical shape. Figure 3.2 illustrates the relationship of energy transfer and the cross section. LET Threshold mention the point where errors are first observed on set and LET saturation is the point where errors stop statistically increasing with LET.

3.4 SEEs in FPGAs

Since the transistor feature sizes have scaled down, the critical charges for SEEs in FPGAs scaled down as well. In order to analyze SEEs on FPGAs, it is necessary to clarify where they can occur in the device first. Every device has different error responses. In order to cope with the error, it is needed to understand the differences and design appropriately.

The upsets experienced by FPGAs fall into three main categories based on the effect of event in the system.

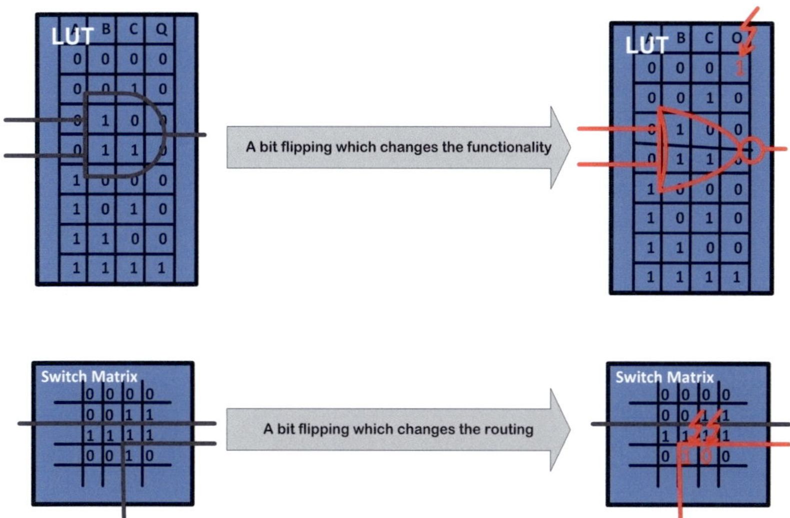

Figure 3.4: A Bit Flip in LUT or in Switch Matrix configuration can change the logic or the routing path and accordingly effect the whole design.

1. resulting in bit flipping in configuration bit stream.
2. affecting user logics.
3. affecting hidden logic or clock tree.

Knowing these categories, these can also be used for SEU characterization and rate predictions. In the following each category of SEU susceptibilities is shortly described.

3.4.1 Configuration bit stream

As described in the previous chapter, in SRAM-based FPGAs, the memory in configured to control the logic implemented by the FPGA device and the routing.

Antifuse-based configurations are SEU immune and SRAM-based ones are SEU susceptible and therefore in Antifuse-based configuration, the SEU error generation modeling is free of the first term (configuration memory bit flipping).

SRAM-based FPGAs, in contrary, use a large array of SRAM memory cells to store the hardware configuration of the device. Typical SRAM cells, and thus the configuration of the FPGA, are especially susceptible to SEUs. [99].

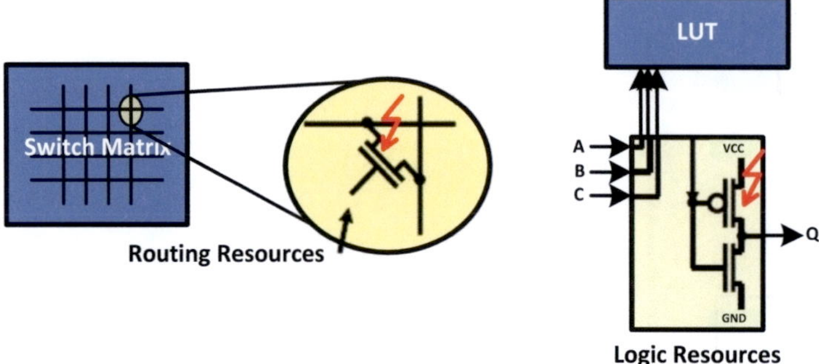

Figure 3.5: User Logic Upset. Upsets can change the status of a transistor in either LUT or Switch matrix and effect the whole design as a result.

Figure 3.4 depicts how an upset can flip a bit in a look up table(LUT) in the FPGA and change the function which is implemented on it. Bit flipping can also cause an upset in the routing matrix which controls the connection between the logic blocks. Such upsets can result in disconnecting routes, creating new routes or bridging two routes. In addition to these effects, upsets in clocking logic (clock tree) can also result in drastic situations and effectively turn off the entire FPGA design [99].

3.4.2 User logic

The user logic includes all user defined logics including flipfops, RAMs, functional logics and the like. The user logic forms the design which is mapped on an FPGA. Both sequential and combinational logic are likely to be affected by SEUs and SETs.

SETs can affect the combinational logics and also squential logic if a glitch is captured. This depends on the design frequency. SEUs change the state of the sequential logic until next cycle of enabled input. Next state capture can be frequency dependent.

In sequential logics, flipflops are connected through a clock tree. If the clock frequency is defined as f_s, then the clock period will be τ_{clk}. Flipflops are the major points of synchronization in sequential circuits in a design. Combinational logics are located between flipflops. Delays come from combinational logics between flipflops (τ_{dly}). It is computed from start point flipflops to endpoint flipflops in general.

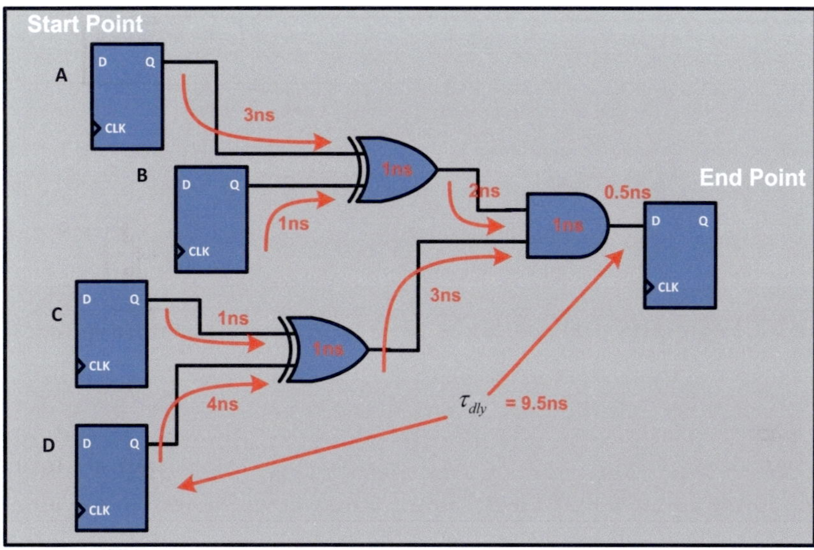

Figure 3.6: Delay on combinatorial logic is significant in propagating or not propagating the SETs on the end point flipflops.

When τ_{dly} is less than τ_{clk}, the system works properly. When an SET occurs in the logic between start and end point flipflops, it takes till $\tau_{clk} - \tau_{dly}$ time to be captured in the endpoint flipflops. Some of SETs are ignorable due to this delay and will not result in an SEU in corresponding flipflop (Figure 3.6) [23].

SET generation occurs when an upset causes an OFF gate turning ON. There is a push-pull between the ON gate and the OFF one which collects charge. P_{gen}, which is the probability of SET generation, depends on amount of collected charge, the strength of gate load, the strength of its complimentary ON gate and the dissipation strength of the process.

In order for the data path SET to become an upset it must propagate and be captured by its Endpoint flipflop. P_{prop} (the probability of SET propagation) only pertains to capacitance of the path, like combinational logic and routing in the path. Small SETs or the paths which have high capacitance have low P_{prop}. P_{prop} contributes to the non linearity of $P(fs)_{SET \rightarrow SEU}$ because of the variation in the path capacitance [23].

P_{logic} is the probability that a SET can logically propagate through a cone of logic. Based on the structure of the combinational logic gates and their potential masking. In Figure 3.7 for example, the AND gate reduces the probability that an SET will logically propagate.

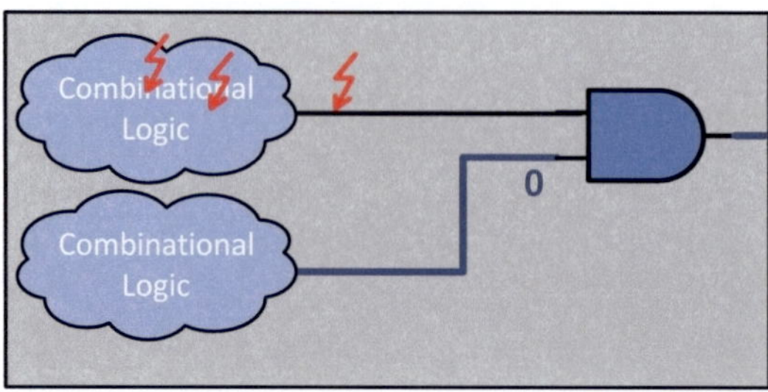

Figure 3.7: Logic error masking can be done based on the behavior of logic gates.

In addition to the above probabilities, the transient width (τ_{width}) is a fraction of clock period (τ_{clk}). Therefore, the probability of capturing an SET at endpoint flipflop is proportional to the width of the transient width and clock width:

$$P(fs)_{SET \rightarrow SEU} \propto \frac{\tau_{width}}{\tau_{clk}} [10] \tag{3.3}$$

The general data path model and combinational logic SETs can be shown as follows [10]:

$$\sum_{i=1}^{\#CombinationalCells} P(fs)_{SET \rightarrow SEU(i)} \propto \sum_{i=1}^{\#CombinationalCells} P_{gen(i)} P_{prop(i)} P_{logic(i)} \frac{\tau_{width}}{\tau_{clk}} \tag{3.4}$$

This means that, transient portion of σ_{SEU} has a direct relation to the number of combinational logic gates and the operational frequency.

In addition to the SETs which affect combinational logic and may be propagated to flipflops to generate SEUs, SEUs can also occur directly in flipflops and change the state of the system. Therefore, for each flipflop, the probability of a failure by SEUs is defined as a sum of the probability of SETs which are propagated and make an SEU in a flipflop and the probability of SEUs which directly occur in a flipflop. $P(fs)_{functional logic}$ for each flipflop can be expressed as:

$$\exists Flipflop \left(\sum_{j=1}^{\#StartpointFFs} P(fs)_{FFSEU \rightarrow SEU(j)} + \sum_{i=1}^{\#CombinationalCells} P(fs)_{SET \rightarrow SEU(i)} \right)$$

However, the SEUs as a result of SETs are more probable in a design based on experiments in [48].

Table 3.1: Orbit average static SEU rates in different durations [94].

Orbit	Altitude (km) Inclination	Solar Max	Worst Week	Worst Day	Peak 5-Min.
		μ (SEUs/Device/s)			
GEO	35, 786 $0°$	$1.6E{-}5$	$1.7E{-}2$	$8.8E{-}2$	$3.3E{-}1$
GPS	20, 200 $55°$	$1.6E{-}5$	$1.5E{-}2$	$7.6E{-}2$	$2.9E{-}1$
Mol.	39, 305/1, 507 $63.2°$	$7.9E{-}5$	$1.6E{-}2$	$8.2E{-}2$	$3.1E{-}1$
Polar	833 $98.7°$	$5.9E{-}5$	$3.5E{-}3$	$2.1E{-}2$	$7.8E{-}2$
LEO	560 $35.0°$	$2.5E{-}5$	$1.5E{-}6$	$1.1E{-}6$	$4.0E{-}6$

3.4.3 SEE Rates in FPGA

Different particles in space have different speed and energy. The energy ranges from 10^6 eV up to extreme 10^{20} eV. Therefore, real estimation of SEE rates at ground level is impossible or very expensive. However, in smaller scales, some experiment are done for Virtex family FPGAs in order to estimate the rate of SEUs or MBUs [101][100]. Experiments show that, the more powerful the particle is, the more likely are Multiple Bit Upsets (MBUs). Such particles can even cause 3 or 4 Bit Upsets in comparison to Single Bit Upsets depending on MeV of particle.

Measurements are made to understand the frequency of effects of radiations. However, it depends on the type and amount of energy of a particle and the duration of its effect on FPGAs. As a result for proton radiations of 63.3 MeV for four different Virtex FPGAs, more than 94 percents of the effects have been single event upsets and less than 0.01 percents 3 or 4 multiple bit upsets.

[94] focuses on four harsh orbits: geosynchronous (GEO), global positioning system (GPS), Molniya, and Polar. It includes a low Earth orbit (LEO) as its reference point and uses a Xilinx Virtex-4 XQR4VSX55 FPGA to measure the SEU rates in different situations. The average SEU rate for each orbit and solar conditions is shown in Table 3.1.

It is important to keep in mind that the test radiation environments include particles with much less energy than what can be seen in orbit environments or in space. Table 3.2 also shows the percentage of resources which are effected in the existence of previous proton radiation. As can be seen, CLBs and IOBs are the most effected ones.

Table 3.2: Frequency of Upset Events and percent of total events induced by proton radiation by resource type for four Xilinx FPGAs [101].

Family	BRAM Events	BRAMi Events	IOB Events	CLB Events
Virtex	0 (0.0%)	in CLBs	8 (2.7%)	286 (97.3%)
Virtex-II	162 (4.0%)	1230 (30.4%)	252 (6.2%)	2404 (59.4%)
Virtex-II Pro	4 (2.9%)	15 (10.9%)	4 (2.9%)	115 (83.3%)
Virtex-4	640 (13.7%)	Unkwn	673 (14.4%)	3,246 (69.4%)

Table 3.3: Frequency of Single Event Upsets and Multiple ones and percent of total events induced by proton radiation for four Xilinx FPGAs [101]

Family	Total Events	1-Bit Events	2-Bit Events	3-Bit Events	4-Bit Events
Virtex	241,166	241,070 (99.96%)	96 (0.04%)	0 (0%)	0 (0%)
Virtex-II	541,823	523,280 (98.42%)	6,293 (1.16%)	56 (0.01%)	3 (0.001%)
Virtex-II Pro	10,430	10,292 (98.68%)	136 (1.30%)	2 (0.02%)	0 (0%)
Virtex-4	152,577	147,902 (96.44%)	4,567 (2.99%)	78 (0.05%)	8 (0.005%)

Current results indicate that the focus on SEUs mitigation, especially on logic, user- and configuration-logic as well, seems to be the correct decision. However, MBUs, are on the rise in newer Xilinx Virtex Family. They are expected to be approximately 1%-3% of the upsets induced by 63.3 MeV proton radiations in these series of experiments. Further study on MBUs will indicate whether these results will have an impact on fault mitigation schemes. However, in order to exactly measure the frequency of radiations on device it is needed to clarify, what kind of particles are there, what strike angle they have and how much energy every particle has.

Table 3.3 shows the frequency of upset events and percentage of total events educed by Proton Radiation (63.3 MeV). It shows that the probability of single event upset is always more than 96 percent in different Virtex FPGAs. As it is shown, the probability of MBUs in 2 bit forms are much more than 4 bit forms in different Virtex FPGA families.

In the next chapter, current mitigation techniques are introduced and their advantages and disadvantages in the existence of SETs and SEUs are presented.

4 Reducing System Error - Common Test and Mitigation Techniques

Hardening techniques must be employed to mitigate device radiation effects. Ionizing radiations may result in SEEs in the configuration memory (SRAM cells) that may alter the function carried out by the circuit that is implemented on the FPGA. This includes both, configuration and user logic errors. Several researches have been done in order to cope with single event effect in FPGAs.

4.1 Mitigation in FPGA

In order to figure out which mitigation strategy is suitable for a design, some mitigation aspects and goals need to be distinguished.

First of all there is an important difference between error masking and error correction. When an error is masked, it will not propagate through the design. When an upset occurs in a part of the circuit, the effect of it remains in the design and will not be corrected. Error masking can result in error accumulation and cause the system to go down.

In contrast is error correction. Here, the error is corrected by using a feedback. Error correction must determine what is right or the correct result to be restored. For instance, if there is a voting system which decides about the correct value, this result can be used as feedback and written back to the error. In this way, the error will be corrected. However, if the voter is not used as feedback, it still eliminates error propagations through the circuit but does not correct it. Therefore it has solely masked the error. To choose the best strategy, an analysis of the circuit requirements is needed and then a decision about masking or correction has to be made. Normally a combination of both is used in big designs.

Mitigation techniques can be embedded into the device but normally they are user inserted. In this case, the user inserts a new mentor component to the system to diagnose and mask(correct) the error. These techniques in FPGAs are divided to two categories: Redundancy based techniques and reconfiguration based techniques.

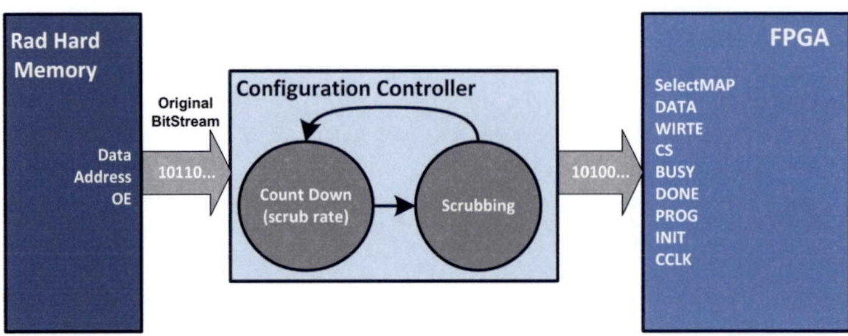

Figure 4.1: Blind scrubbing, which includes periodically reconfiguring the FPGA

One of the most important reasons to use a mitigation technique is its simplicity. Simple implementation is important because any incorrection in its implementation may help raising the errors in the design[23]. However, in order to consider the mitigation strategy some major aspects such as the upset rate and the criticality of the components which are implemented on FPGA must be taken into accout.

4.2 Reconfiguration based techniques

A countermeasure, which is known as *scrubbing*, is to write the original configuration to the memory periodically hence correcting flipped bits. Scrubbing can be described as the process of reloading the configuration bitstream, so upsets in configuration are corrected [24]. Scrubbing restores the desired reconfiguration periodically regardless of any SEU fault on device. It can be done in two manners: blind and readback [31][68].

4.2.1 Blind Scrubbing

Blind scrubbing is simpler than readback. Here, the scrubbing is done periodically after a downtime. Due to degrading overall performance which is the result of periodically stopping the system operation and reconfiguring it, some FPGAs like Xilinx devices support a model which is called active reconfiguration. This allows concurrent system operation and in parallel reading/writing of configuration bits.

The frequency of such a process might be determined based on an expected upset rate (number of upsets per second) or simply based on the fastest possible reconfiguration speed for the device (bits per second).

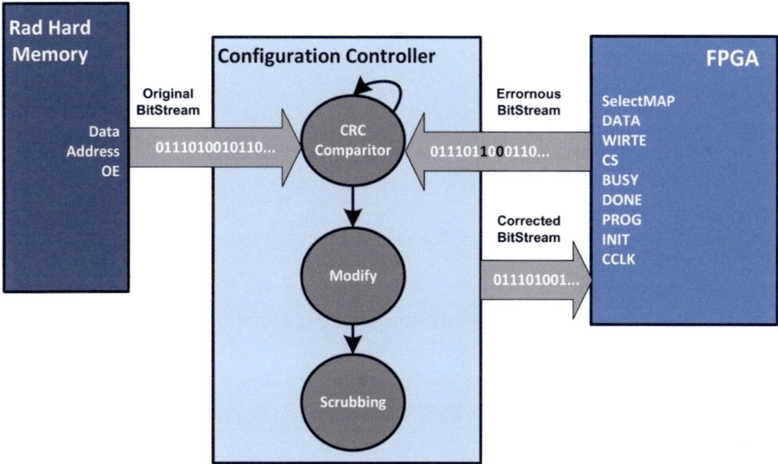

Figure 4.2: Readback Scrubber checks CRCs and compare them to reconfigure the FPGA when bitstream is errornous

Faster reconfiguration is desirable since it minimizes the time during which output errors might persist, but continuous configuration scrubbing consumes additional power. The minimum scrubbing cycle duration is determined by the speed of reading/writing configuration bits and the size of bitstream (which depends on the size of the circuit), to be scrubbed. In Figure 4.1, an abstraction of blind scrubbing is shown.

4.2.2 Readback Scrubbing

In readback scrubbing, there is a detector which only refreshes the configuration memory contents when either sensitive bit upsets or a certain number of non sensitive upsets are detected. This involves both reading and reloading the content. More sophisticated scrubbing techniques involve fault isolation to determine which specific portions of the chip require repair. Scrubbing by reading CRCs is shown in Figure 4.2.

A feature of modern SRAM-based FPGAs is dynamic partial reconfiguration. Partial reconfiguration allows reconfiguration of a smaller portion of design. In this way, when a misbehaviour is detected in the design, it is not needed to reconfigure the whole FPGA, but rather the defected part by using partial reconfiguration [31]. The partial reconfiguration can especially optimize the scrubbing because in this case scrubbing can be done online while the device is performing its task[32].

Figure 4.3: Probability of i upsets per scrub cycle for a Xilinx Virtex 4 FPGA in solar max condition in GEO with 15 ms scrub period[94]

In general, the frequency of configuration is called scrub rate. It depends on several issues such as the SEU rate and the other mitigation techniques which are used.

4.2.3 Failures in Scrub Cycle

[94] computed the probability of i upsets during a single scrub cycle which is $P(A_i)$ and is modeled with a Poisson distribution as shown in Equation 4.1:

$$P(A_i) = e^{-v}\frac{v^i}{i!} \tag{4.1}$$

which v is the average number of SEUs per scrub period and is calculated by multiplying the orbit average uspset rate, μ, by the scrub period, t_s as follows:

$$v = \mu \times t_s \tag{4.2}$$

Using the information of Table 3.1, [94] analyzes the $P(A_i)$ for a Xilinx Virtex-4 FPGA as shown in Figure 4.3.

As results show, for critical applications, scrubbing can not be the only method which is used to cope with failures. Redundancy based techniques which detect, mask, and correct errors, are also essential in order to cope with high single event rates. However, a combination of scrubbing and redundancy based techniques

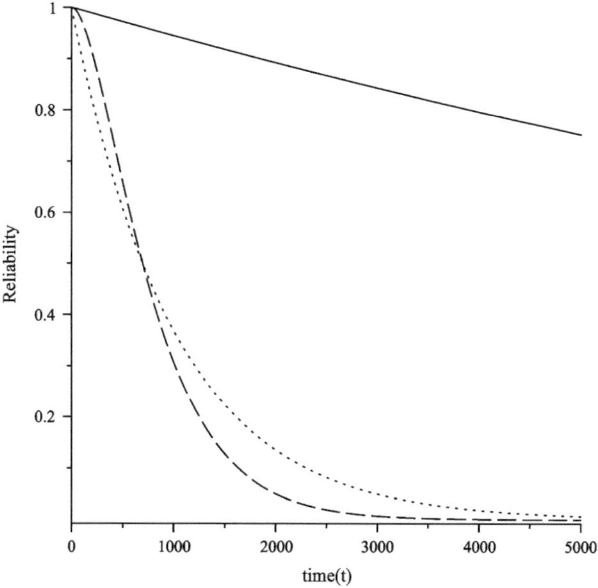

Figure 4.4: Reliability of a system with NO TMR, TMR, and TMR with scrubbing [38].

is the ideal strategy to cope with SEUs. [38] and [24] consider this issue.

However, as in Figure 4.4 can be seen, even Triple Modular Redundancy (TMR) without scrubbing can not improve the reliability of the system. Redundancy based techniques, are the focus of this work.

4.3 Redundancy based techniques

Hardware redundancy techniques need additional hardware components for masking the presence of SEUs or correcting them [82]. In the case of FPGAs, fault detection and masking can be achieved by triplicating the circuit which is implemented on FPGA [33][80] [89].

Triple Modular Redundancy (TMR) is a well known fault tolerant technique for tolerating against the errors in integrated circuits. Traditional definition of TMR mentions that, the TMR scheme uses three identical logic blocks performing the same task. The corresponding outputs are compared through majority voters. The simple scheme is shown in Figure 4.5.

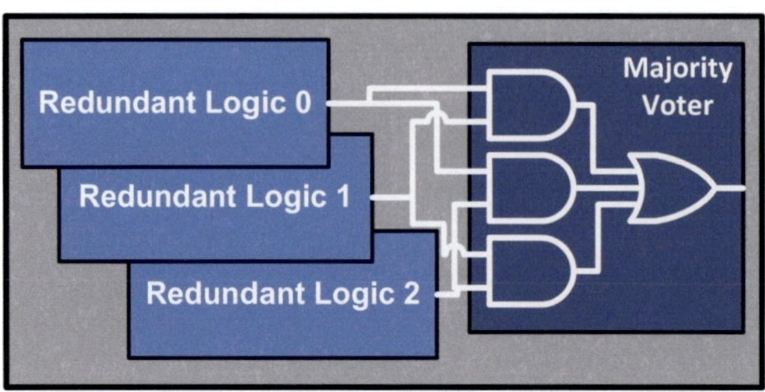

Figure 4.5: Triple Module Redundancy with majority vote

Implementing TMR to prevent the effects of SEUs in technologies such as ASICs is generally limited to protecting only the memory elements since combinational logic and interconnections are less sensitive to SEUs. When the configuration memory of FPGAs is considered, the TMR implementation is different than ASICs since a modification in the configuration memory may affect every FPGA's resources including combinational and sequential resources like is shown in Figure 4.6. This means that, the whole design can be affected by upsets and in the simplest form, three copies of the whole circuit, including I/O logic, must be implemented to harden it against upsets [123].

Redundancy based and reconfiguration based techniques can also work together using dynamic partial reconfiguration. For example in a TMR design, if one of the copies is affected by an SEU, one solution is to dynamically reconfigure this part. Normally, if errors accumulate in TMR, the whole design must be reconfigured which is a big overhead. However, by using dynamic partial reconfiguration, the affected copy will be reconfigured while the other copies are still running [19][96].

Single voting is a single point of failure while an SEU affects it [33]. If the corresponding SEU hits the voter, then TMR might not function anymore. Due to this fact, normally TMR is implemented by triplicating the voters.

TMR, has been developed and improved through years of research. For ground based complex systems, TMR might be practical while it corrects single failures. If it is assumed that errors upon configuration or user logic functions can occur one at a time, then TMR is able to mask them very well and prevent any corruption at the output.

It is necessary to notice that, in high upset rates, if there is a certain probability for upset for one of the copies, by triplication, the probability will be also triplicated. It means that, in high upset rates, redundancy will not necessarily result in more reliability, if it is not accurately selected.

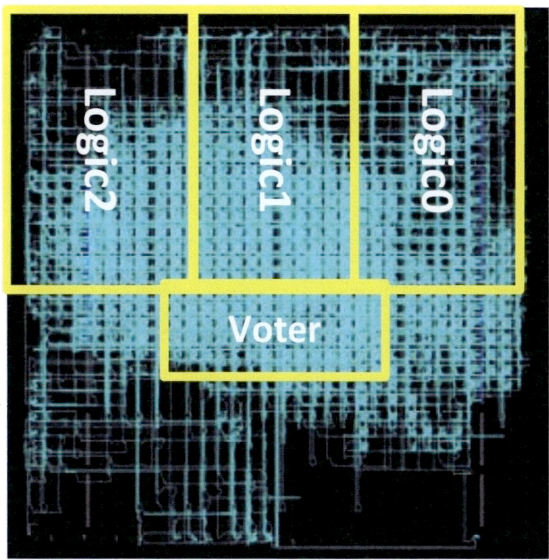

Figure 4.6: Triple Module Redundancy abstraction in FPGAs. Three copies of a design in addition with the voter are mapped on FPGA

For example, in space, upset rates are high and there is more possibility for multiple errors. If the conventional TMR is selected as the mitigation technique, it might result in corruption. If two upsets occur simultaneously in two domains of the same triplicated logic, the majority vote would choose the corrupted result instead of the correct one. In addition, if one copy goes down, the masking capability of the voter is lost. The way to cope with this is that, the device which goes down must be corrected before any other upset can occur. However, it might not be possible in critical applications [23].

One of the major disadvantages of TMR in high upset rates is that, the whole design is triplicated and then a voter decides about the correct output. In this way, there is no internal accessibility to the design. If one error occurs in one of the copies, the copy will go down. If more than one error occurs, also the copy will go down. There is no difference between one or more errors, because there is no accessibility to the inside.

An attractive alternative is to localize the TMR in the design. In this way, every part of the design will be locally TMRed and voted. This results in local error masking and improves the reliability in case of more errors. In this work, this is generally named Fine Grain TMR or FGTMR. This work focuses on finding suitable fine grain techniques to make a design high reliable.

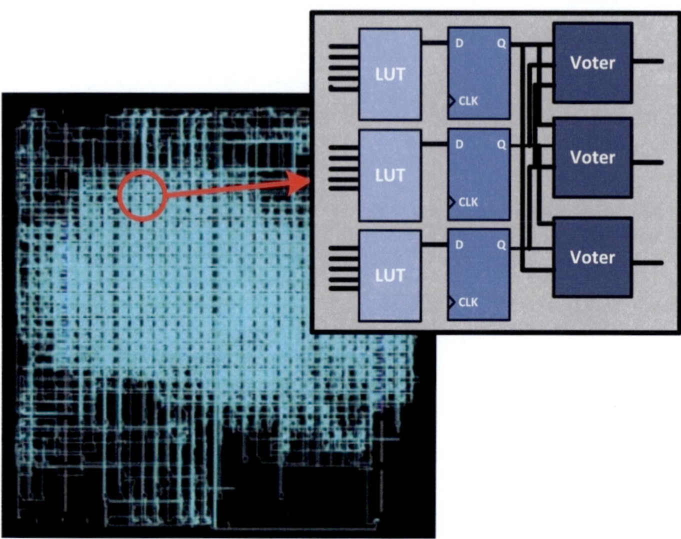

Figure 4.7: Using Fine Grain View to TMR in homogeneous FPGA architecture

4.4 Fine Grain TMR - FGTMR

Although aforementioned TMR redundancy methodology makes some systems fault tolerant, there are scenarios where these approaches can fail. One of the obstacles of TMR in its simple form, is its inability to cope with more frequent upsets in the design. If parts of the combinational logic or one of the flipflops is affected, the whole copy will go down. In addition, when an error changes the state of a flipflop, it may remain there till it is rewritten again.

In this work, conventional TMR is named as Coarse Grain TMR (CGTMR). CGTMR is a view to the fault tolerance issue, which does not cope with the fine grain parts of the system. In other words, a design is triplicated and when an upset occurs, it is not clear where exactly in the design is affected. This is specially an obstacle when more than one copy is affected in CGTMR which is often in case of multiple SETs.

A coarse grain view to redundancy might not be sufficient for scenarios with high rates of SETs, SEUs and MBUs. If the system would be able to detect failures and upsets locally, it can be able to deal with high failure rates. These challenges lead to a more localized view regarding the application of fault tolerance methods. It is categorized as Fine Grain TMR or FGTMR.

The principle behind using FGTMR is that, in a homogeneous architecture like FPGAs, TMR can be applied to the fine grain homogeneous parts of the design

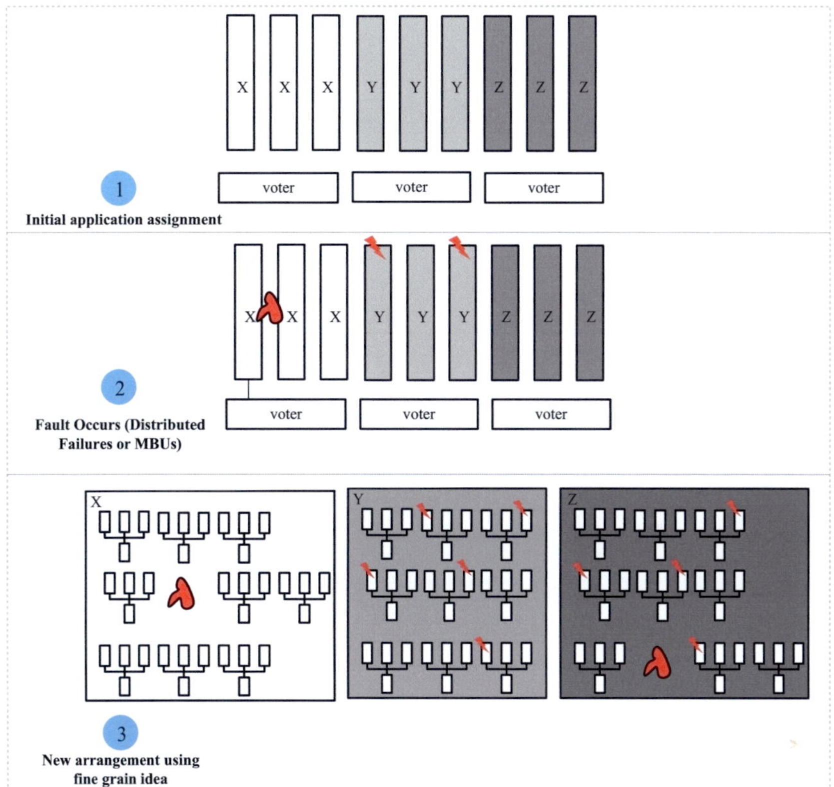

Figure 4.8: CGTMR and FGTMR comparison. In the second scenario, an MBU occurs for design X and two parallel SEUs occur for design Y. In both cases the voter will be unable to decide for the correct value while more than one design is corrupted. In the third scenario which is FGTMR, designs can tolerate against more number of upsets.

instead of the whole design (Figure 4.7). Homogeneous parts of FPGAs include combinational logic and flipflops which in FGTMR are locally triplicated and using local voters mask and correct errors.

The level of granularity can be determined based on the requirements of the system. Failures are either distributed or are focused in a special area of architecture. In both cases, FGTMR can be useful. In Figure 4.8, CGTMR and FGTMR are compared in a design.

4.5 Estimating reliability on different TMR granularities

A simple reliability probability modeling can be used in order to analyse finer granularities in TMR. Different granularities can be modeled and triplicated to find a relationship between granularity level and reliability of design. Results of the analysis show when the granularity level is increased the reliability is improved. Here the reliabilities of two different levels is analyzed and discussed.

4.5.1 Coarse grain level redundancy

Assume that circuit A is triplicated in coarse grain. When a copy is affected by failures it will perhaps generate the incorrect output and due to this will be assumed out of order. $P_{A_1}(0)$ is defined as the probability of one copy, that has no failure inside. Assume that a coarse grain A consists of m fine grain components. The probability of A to work correctly is defined as $P_{A_1}(0)$ which means that, there is 0 failures in m fine grains. It will be defined as p^n. The other copies of application have exact the same probability structure. The probability model in application level is then given by:

$$\begin{aligned} P_{\text{Reliability}_A} = \; & P_{A_1}(0)P_{A_2}(0)P_{A_3}(0) \\ & + (1 - P_{A_1})(0)P_{A_2}(0)P_{A_3}(0) \\ & + P_{A_1}(0)(1 - P_{A_2})(0)P_{A_3}(0) \\ & + P_{A_1}(0)P_{A_2}(0)(1 - P_{A_3}(0)) \end{aligned} \tag{4.3}$$

which can be briefed as follow:

$$P_{\text{Reliability}_A} = p^{3n} + 3 \times (1 - p^n) \times p^{2n} \tag{4.4}$$

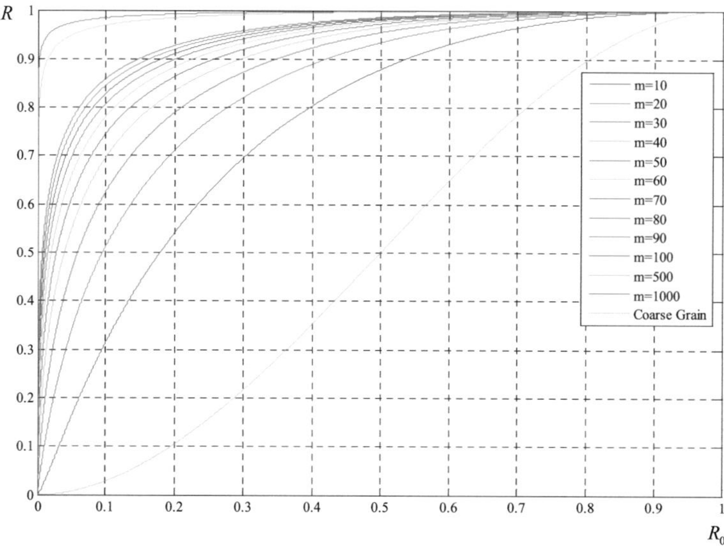

Figure 4.9: Reliability of Coarse Grain TMR vs Fine Grain one with different number of grains(m)

4.5.2 Fine grain level redundancy

The probability model in a fine grain view to redundancy is defined. Assume, p is the probability of one grain to work. If R_0 is defined as the probability of the whole m fine grains to work, then p will be $\sqrt{R_0}$. The probability of TMR in a fine grain M then will be defined as: R_M as $R_M^3 + 3 \times R_M (1 - R_M)$ so $3 \times R_M^2 - 2 \times R_M^3$ and for the whole design with m blocks there will be:

$$P_{\text{Reliability}_{R_0,m}} = (3 \times R_0^{2/m} - 2 \times R_0^{3/m})^m \tag{4.5}$$

for different m and also in coarse grain case, the experiment has done and the analytical results are shown in Figure 4.9.

As can be seen in Figure 4.9 the reliability probability of different fine grains is considerably in better condition than coarse grain one.

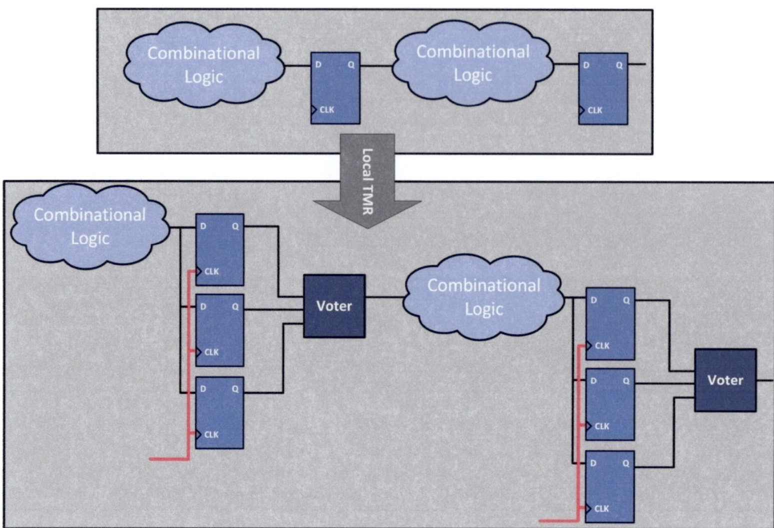

Figure 4.10: A design after applying Local TMR

4.6 Different Granular Techniques on FPGAs

Different strategies can be selected in order to apply FGTMR to a design on FPGA. Here different techniques and their disadvantages have been described in detail.

4.6.1 Local TMR - LTMR

In order to mask and correct SEUs which occur in fliplflops, one solution is to triplicate flipflops and add voters to the output of these triplicated flipflops. As Figure 4.10 shows, the datapath remains a singular path and redundancy is done just in sequential level at the flipflops. The correct voted value can be transferred back to every flipflop to correct them. This eliminates any error accumulation on flipflops [23].

This method provides mitigation by redundancy, voting and feedback. What is not protected here is clock and reset and also transient upsets (SETs) in data path. Looking at the SEU resources from the previous section (Equation 3.3), the most dominant factor in LTMR will be SETs which potentially can result in SEUs. The SEUs which occur directly in flipflops, are masked by majority voting and are corrected by feedback.

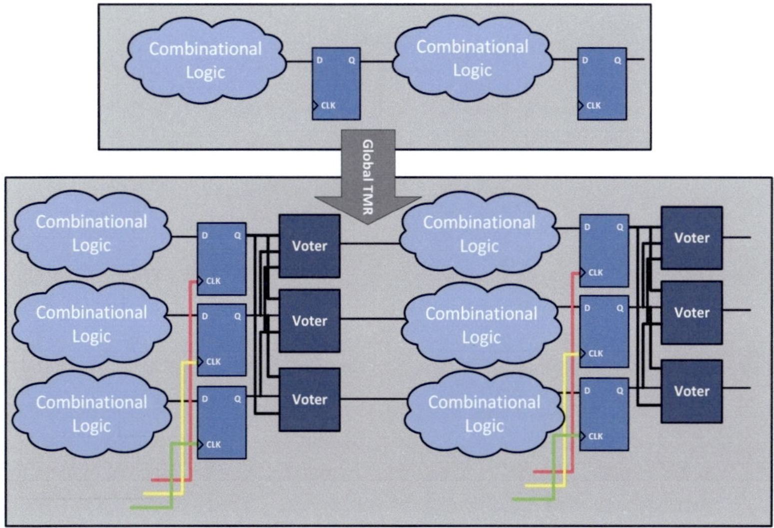

Figure 4.11: Global TMR

Generally, SETs can occur in parallel in different parts of the data paths. If one of SETs is propagated in a flipflop, it is possible to be masked by majority voting. When more SETs are propagated, the voter will be unable to recognize the correct result. To cope with SETs in the datapath, extra redundancy in the datapath is needed.

Every time that multiple SETs occurred in the datapath, it can easily affect more than one copy and make the whole mitigation strategy fail. As previously mentioned SETs are potential reasons of SEUs which result in in-correction in the whole system. MBUs can also effect more than one copy at the same time and make the fault tolerance strategy fail.

In addition to the disadvantages of voters, all flipflops share the same data path and every SET in these data paths can affect them. Global routing can also be a point of upset, which can affect more than one copy at the same time and cause SEUs. However, more than one SEU in the same time at two different copies of flipflops is not protectable by LTMR (Figure 4.12).

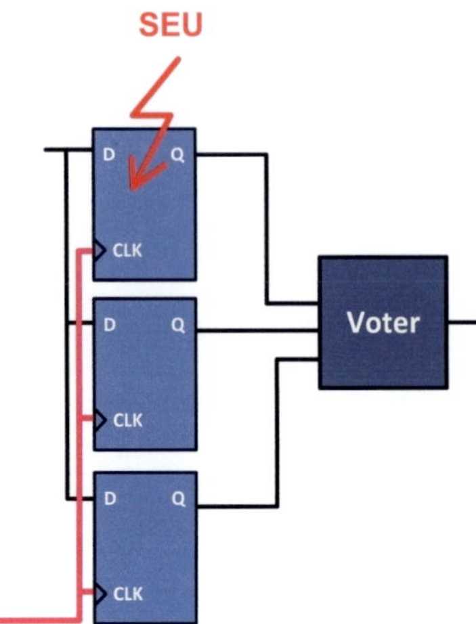

Figure 4.12: a SET can result in SEU by affecting a flipflop or propagating through it and changing its state. In LTMR there is no feedback to FFs to correct the state and multiple SEUs can result in system failures.

4.6.2 Global TMR - GTMR

An extension to LTMR results in GTMR. Global TMR or GTMR triplicates all parts of the design including combinational logic, and flipflops. There are three different clocks and three different domains. SEUs and SETs are protected by triplicating and majority voting. This approach has shown in Figure 4.11. Voters can write the correct output back to the flipflops in order to avoid error accumulation [18].

Although GTMR can tolerate against SETs and SEUs, it is very power and area hungry in comparison to LTMR.

In addition, in GTMR, not only domain placement is still a problem but clock skew can also be a concern in GTMR because of three different clock domains. If clock skew is larger than the feedback, timing correction does not work properly any more. This cannot be diagnosed due to complexity reasons. It is possible that the system is not able to mitigate as it was expected.

4.6.3 Distributed TMR - DTMR

A simplification of GTMR is Distributed TMR or DTMR. Here, every thing is triplicated except for the global clock routing and reset. This protects the data

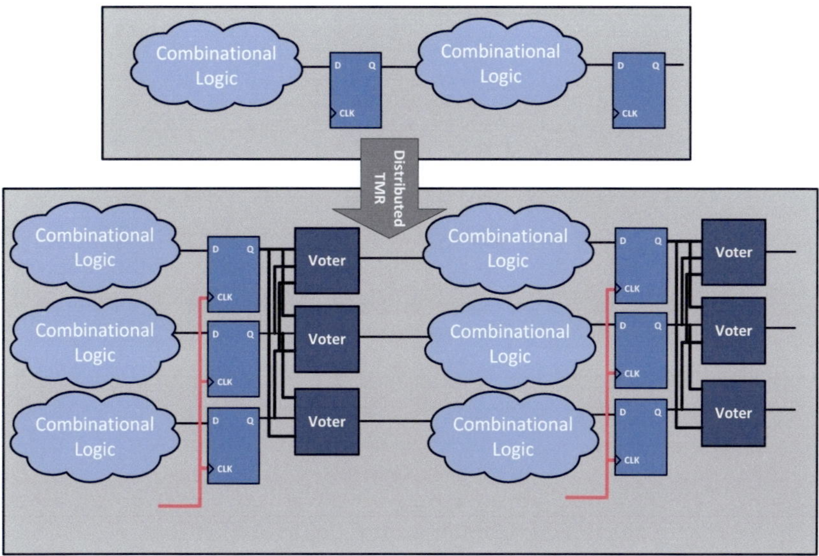

Figure 4.13: Distributed TMR

path against SETs. However, it is not clear if two SETs at the same time effect two copies and propagate in corresponding flipflops, how the voter recognize the correct output. Figure 4.13 presents DTMR scheme.

In DTMR there are still voter problems like in GTMR and LTMR, but the clock skew problem is not a matter any more. On the other hand, SEUs in clock routings can be a problem which is not fixed because clock remains singular.

What is not discussed is that, when complex FPGAs are used, there are generally elements which can share routing matrix. Strikes on shared routing matrix can take out two domains and this causes the mitigation to fail.

One idea is to place the elements far from each other. However, this is a trade off. Because if elements are placed far from each other, they need more nets and probably make the critical path longer.

In all of the TMR techniques, verifying the TMR by making sure that every node has been triplicated, always could be a problem. There is no fine grain control on the system and every susceptibility in finer grains, such as two SETs in one domain or in two copies of one fine grain, can result in mitigation technique failure.

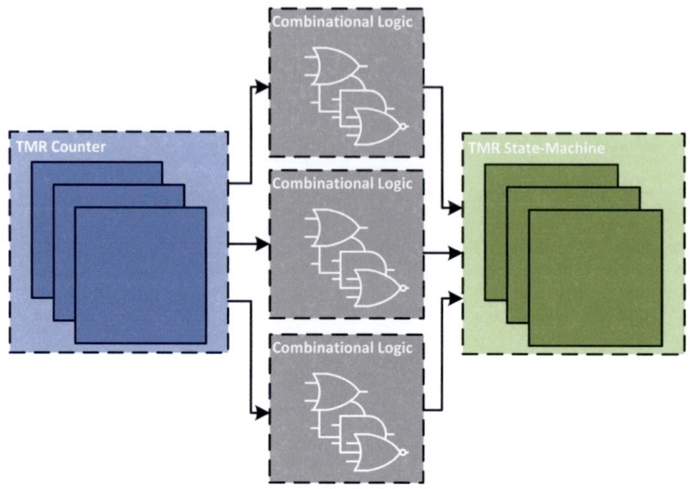

Figure 4.14: Combinatorial Logic in XTMR

4.7 Integrating FGTMR in FPGAs

In FPGAs, TMR insertion is based on homogeneity of reconfigurable architectures, respectively SRAM-based FPGAs. By focusing on architectural design primitives it is possible to define localized redundant patterns for LUTs and apply these patterns to any design that is mapped to the FPGA. This allows automatic introduction of TMR as sole exchange of modules based on a library approach(e.g. LUT vs. TMR-LUT)(Figure 4.7).

Because all FGTMRs are a small TMR system each of them can deal with one single failure without disturbance of the overall system. The probability of multiple failure occurrence in two redundant fine grain is much less than the coarse grain one. Hence the fine grain redundancy is expected to be a more tolerant structure against multiple distributed failures.

4.7.1 Xilinx TMR - XTMR

Xilinx uses the fine grain redundancy in the form of GTMR and has developed the TMR tool for Virtex FPGA families. In XTMR, the voters are also triplicated, which consumes more logic but improves the reliability. Outputs are further enhanced with a minority voter to prevent drive tights which may damage device and/or board components. When the inputs and outputs are also triplicated, the SEEs are eliminated.

In order to correctly implement the TMR circuit on Virtex architecture, XTMR groups the data structures in a design into four different types and applies the FGTMR depending on the type of the structure to avoid in-correction. The four different types include: Throughput Logics, State Machine Logic, I/O logic and special features like Block RAMs. Here it is shortly described the way that FGTMR is applied to each type.

4.7.1.1 TMR in Throughput Logic

In order to implement TMR for throughput Logic, XTMR simply creates three copies of the basic structure. This includes three versions of inputs and outputs of throughput logic. However, throughput logics are naturally between various state machines in a design. XTMR simplify the TMR implementation by creating hierarchical boundaries around individual state machine logic structures. Figure 4.14 shows the throughput logic between state machines which is simply triplicated. There is no voter at the end of throughput logic because it is constructed inside the state machines.

The point in throughput Logic triplicating is that, it is not shared between two state machines. Any soft error can propagate through the next state machine and will not get caught in any loop. As in [33] is mentioned, the only purpose for the redundant logic between state machines without voters is to carry the triple redundant output of the previous state machine without creating a single point of failure. Not using the voters at the output of redundant logics, eliminate the interconnection between the three outputs.

4.7.1.2 State Machines

State machines are by their definition state dependent. If they are simply triplicated and voted without feedback, it is likely that the voting fails to recognize the correct value after a while due to SEUs. Feedback will solve this problem. Figure 4.15 shows the problem of voting in case of one bit Counter and Figure 4.16 the solution using feedback paths after redundant voters.

Therefore, the basic concept of XTMR includes, triplicating every state machine and combinatorial logic and inserting majority voter at the end of every loop or feedback path like DTMR. The use of three redundant majority voter eliminates the single point of failure and provides three logic path inputs and outputs [33].

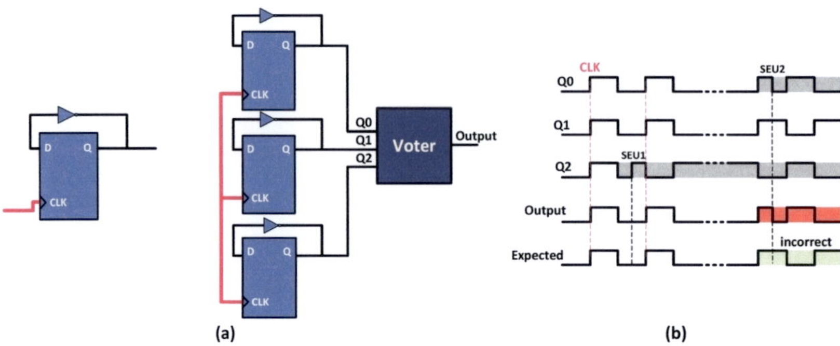

Figure 4.15: (a) Single Voter One bit Counter TMR without feedback. (b) Problem after sequential SEUs.[33]

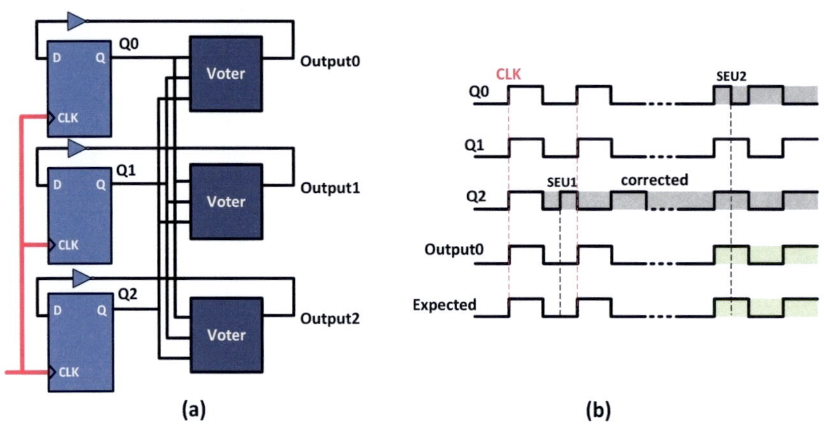

Figure 4.16: (a) Distributed TMR which is applied to state machines in XTMR. (b) the waveform is corrected. [33]

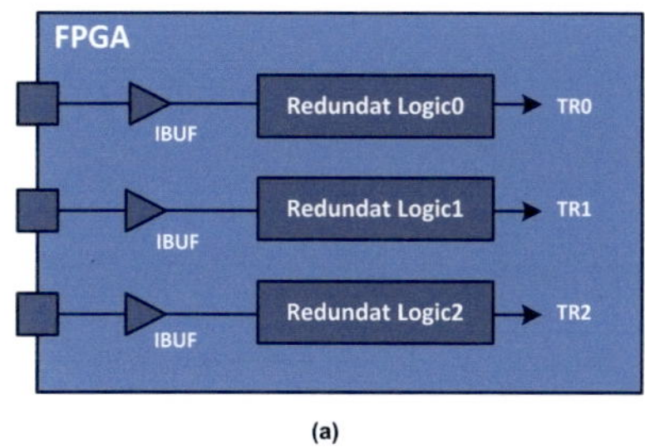

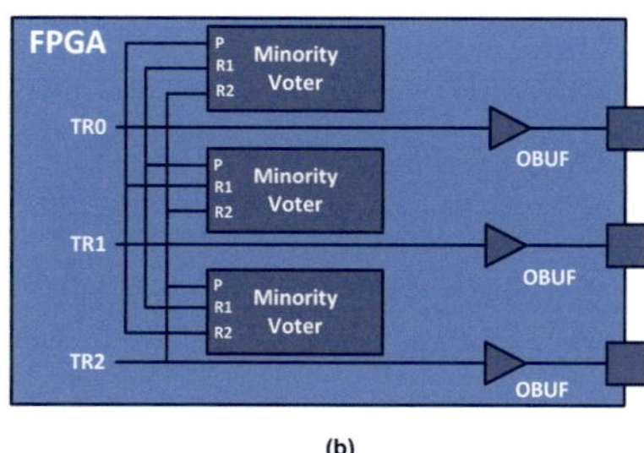

Figure 4.17: (a) Triple Redundant FPGA Inputs (b) Minority Voted TMR FPGA Outputs. [33]

4.7.1.3 I/O Logic

The main purpose of XTMR is removing the single point of failures from the design. This can begin with the FPGAs inputs. Every input failure can be a starting point for propagating an error in the whole design if it is not detected and recovered. Because of this reason, every input is triplicated and if one input suffers a failure, it will only affect one redundancy. However, this means a huge amount of I/O resource usage and must take into account when choosing a device.

As well as inputs, outputs are also triplicated in XTMR. Since there are three copies of the logic path, if there is just a single path which brings them to output, this single path can be critically a single point of failure. Every redundant copy sends its output to a minority voter. The minority voter (Figure 4.18) feeds the output buffers which are hardwired together on the circuit board.

The structure of minority voters is as follows: Every minority voter has a primary path and two secondary paths. If the primary path is a part of majority paths(the other two secondary paths) minority voter allows it to feed the output buffer and not vice versa. The structure of I/Os in XTMR is presented in Figure 4.17.

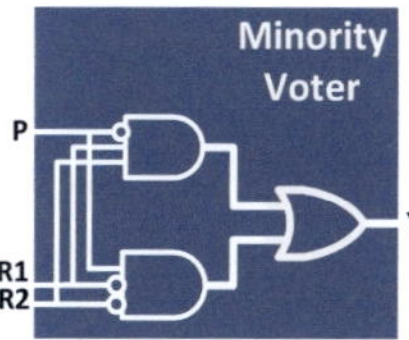

Figure 4.18: Minority Voter Circuit

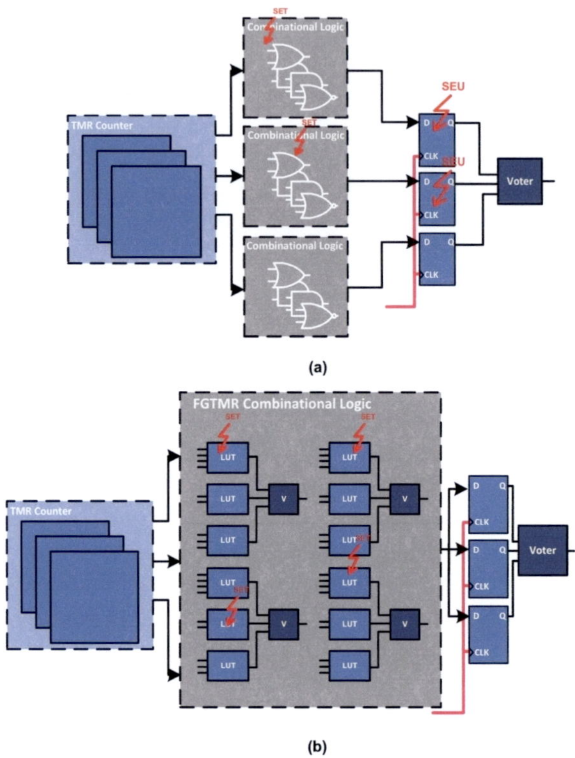

Figure 4.19: (a) XTMR will fail when more than one combinatorial logic copies is affected by SET and propagate it to the corresponding FF. (b) FGTMR approach solve this problem by fine grain FGTMR.

4.7.1.4 Special Architecture Features

In XTMR target FPGAs which are the Virtex Family, there are some more special feature architectures like BRAMs, DLLs, DSP, and etc., which make the design more efficient and allow higher performance implementations. However, XTMR is still under development and as [33] mentions, currently this components are used as a triplicated form in order to be compatible with the other parts of the design. More details on these special architectures can be found in [33].

4.7.2 XTMR Failures

[73] did experiments on possible XTMR failures by inserting faults in the bitstreams. They observed that the main reason of XTMR failures is the undesirable shorts occurring due to in-correct routing as a result of bit flipping in the bitstream or configuration latches. Normally, the bit which is affected makes a connection between two redundant modules. After this kind of bit flipping both modules fail.

The basic idea to avoid this, is logic partitioning. In this way, firstly logics are partitioned and triplicated majority voters are inserted after every stage of minimal logic [122]. Due to this method, the faulty connection can no longer lead to a failure as the voters along this connection will vote out the erroneous stage output. [122] Has measured different logic partitioning levels and counts the number of wrong answer for a digital low pass filter as case study. Results show about 4% of failures in the case of XTMR with minimum partitioning and 0.98% in case of maximum partitioning in logics.

In another research, fault injection mechanisms are used to realize failures in XTMR and it proves that XTMR does not provide 100% protection against SEUs. It is shown that a single SEU is able to produce multiple errors that can lead to TMR failure.

Consequently an algorithm is used to make the XTMR robust and insensitive to SEUs. The algorithm is called the Reliability Oriented place and Route Algorithm (RoRA) ([116]) which places and routes designs according to a set of design rules that make designs insensitive to SEUs. This research focuses on the failures in routing resource instead of CLBs. However, after applying RoRA, routing failures are reduced. But the failures in CLBs and logic parts still remains an open issue [16][117].

This results are a motivation to do research on a fine grain TMR approach which suitably is applicable on FPGAs Look Up Tables and flipflops. The rest of this chapter issues this approach.

4.8 New approach on FGTMR

XTMR is specific to the Virtex FPGAs and not the other FPGA families like Altera SRAM-based FPGAs. Also, it does not offer flexibility in selectively applying it to the design. Therefore, it is often expensive in terms of resource utilizations, power consumptions, etc [33][13].

Moreover even by XTMR, the design is still susceptible to SETs. In XTMR, combinatorial logic is a black box which is triplicated between two state machines.

When multiple SETs occurs in the combinatorial logic, they can potentially propagate through and effect the flipflops in the next state machine. When more than one flipflop is affected, even with feedback, the incorrect result will change their states and redundancy strategy will fail to recover the design.

To reduce the cost of mitigation, one way is to apply it selectively on a subset of the design on FPGA. In this work a FGTMR tool is developed which has the property to cope with some issues on XTMR. The approach of this tool looks to the design on the FPGA as a fine grain of homogeneous LUTs or LUT-FFs pairs. Every fine grain is triplicated and voters are added at the outputs. It is actually the full partitioned XTMR which eliminates affecting the combinatorial logic or probably flipflops with multiple SETs and SEUs and in addition eliminates possible errors which effect more than one copies in TMR.

In addition to multiple SETs, the FGTMR can be applied to a subset of the design which is sensitive to an upset. In [98] some experiments have been conducted to recognize the sensitive parts of a design. Using this experiments, in every design, FGTMR can be applied to the most critical sections of a design. This reduces the area and power usage of the system. By selectively applying mitigation to a design, one can find the most effective balance between FGTMR cost and reliability cost in high upset rates [98].

In Figure 4.19, the idea of fine grain TMR on FPGAs is specified. It is implemented to evaluate the area and fault tolerance by more frequent voting in the fine grain parts of the design without using the XTMR. The FGTMR can be generalized for all kinds of FPGAs based on their homogeneity issue.

4.8.1 Automatic Insertion of FGTMR in Synthesis Flow

In general, the commonly used design flow for FPGAs consists of three phases (see Figure 4.20). In the first phase a synthesizer is used to transform the high level circuit model in hardware description language into a netlist in RTL level. The RTL level design in the second phase is used by technology mapper to be transform to a gate level model which is composed of LUTs, FFs, BRAMs and the like. Finally the physical implementation is achieved by placing and routing the mapped design onto the FPGA.

For current FPGAs like Virtex-5 devices [63] physical placement is already done within the mapping step. The RapidSmith XDL parser tool [103] is used, which convert the NCD form of the map result to an XDL format data structure and makes it easy to apply different modification algorithms after map. The modification algorithm is in this case FGTMR which triplicates every LUT and insert proper voters to the design.

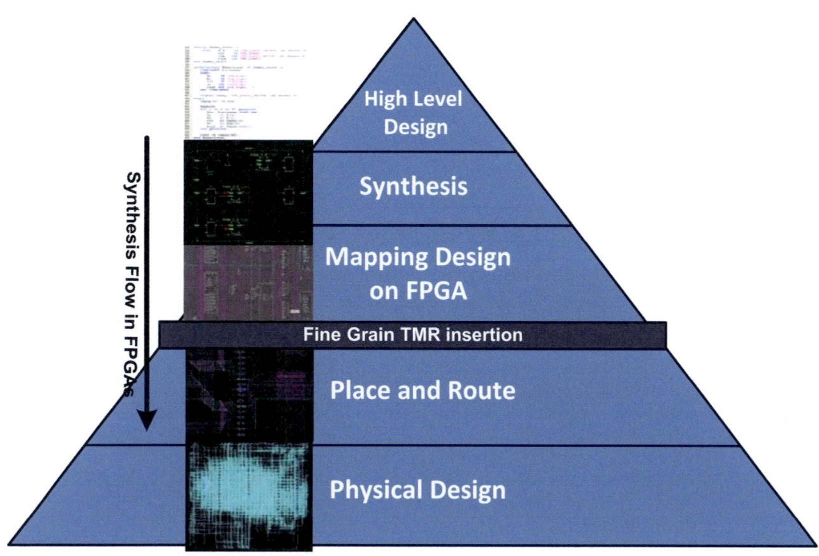

Figure 4.20: Fine Grain TMR in Synthesis flow.

The abstract framework of the automatic fault tolerance tool is depicted in Figure 4.20. Further modifications in Fine Grain fault tolerance are the focus of this work. Due to this, more integrations are done in the synthesis flow for fault tolerance reasons. Fine Grain fault tolerance integrations in synthesis flow are described and detailed in the following chapters.

If FGTMR covers the whole design and not a part of it, the area usage is more than in CGTMR because of triplicating every fine grain part. But in comparison to XTMR, it makes practically no difference while XTMR triplicates the logic as well the pure logic consumption.

Major difference are the voters which are necessary for each fine grain in FGTMR. In XTMR the voter size can be usually neglected because it is extremely small in comparison to the three functional copies. This does not apply to FGTMR because the voter's size is similar to each of the three LUT copies, hence resulting in an overhead of four times (or six times in case of triple voting) of the singular copy instead of three times.

However the results show that this overhead might be worth its price because this methodology can recover from a huge variation of failures from SETs to manufacturing and process variation effects(two completely different resources of failure).

Table 4.1: Area overhead comparison between the original circuit, the circuit which is coarse granular redundant and the circuit which is fine granular redundant.

Parameters	KCPSM3 (PicoBlaze of Xilinx)			3DES (Data Encryptor Standard)		
	Original	CGTMR	FGTMR	Original	CGTMR	FGTMR
Number of BUFGs	1	3	3	1	3	3
Number of External IOBs	58	174	174	190	570	570
Number of Slice Registers	76	228	228	263	789	789
Number used as Flip Flops	76	228	228	263	789	789
Number of Slice LUTS	149	621	846	531	2163	5487
Number of Slice LUT-Flip Flop pairs	153	633	858	642	2100	5595

Table 4.2: Tolerance of design against LUT defecting. C indicates the number of LUTs which are defected concentrated in a coarse grain copy of design while D shows the number distributed broken ones. Fault tolerance percentage of every method is shown in the last row

	KCPSM3 (PicoBlaze of Xilinx)			3DES (Data Encryptor Standard)		
	Original	CGTMR	FGTMR	Original	CGTMR	FGTMR
Number of Slice LUT-Flip Flop pairs	153	633	858	642	2100	5595
Probable Num of broken LUTs	0	153(C)	286(D)	0	642(C)	3145(D)
Fault Tolerance Percentage	0	24	33	0	30	56

4.8.2 Results on Area and Fault Tolerance

The presented methodology has been verified using a Xilinx XC5VLX110T Virtex-5 FPGA [63]. The automatic approach is applied to different circuits and two benchmarks are selected to prove the FGTMR and compare the area overheads. First one is Kcpsm3, PicoBlaze from Xilinx which supports a program up to length 1024 instructions [65], and the other one is 3DES which is a Data Encryption Standard cipher algorithm from OpenCores [93].

To check the area overhead of FGTMR with CGTMR and the original design are compared. Table 4.1 shows the result of original design after Place and Route using Xilinx ISE Synthesis Tool in comparison with Coarse Grain TMR implementation of the circuit and Fine Grain TMR which the automatic tool has inserted in the flow of synthesis tool. As the results show, area overhead is about 2.5 times for pairs of FF and LUTs for FGTMR.

To check the fault tolerance percentage of the FGTMR, LUTs are injected with faults, by changing the initialized data of LUTs after post routing simulation result(VHDL format). Then the design is simulated after this modification and is checked for correctness. Although there is an area overhead for FGTMR, Table 4.2 shows that using FGTMR, the circuit is able to tolerate against 33 percent and 56 percent of the faulty pairs of LUTs and FFs in two benchmarks in comparison with CGTMR which is able against about 30 percent of pairs and just in the cases that failures are concentrated on one coarse grain copy of design. As described in the previous sections, the advantage of FGTMR is multiple failures tolerance which are distributed on the circuit.

Figure 6.21 in Chpater 6 illustrates another possibility for FGTMRs. It is related to so far as possible placement of fine grain LUTs on the design to tolerate against the MBUs, which can effect a special place of the FPGA which contains an amount of slices. If every three redundant LUT be computed to place so far as possible on the design, at least two of them will scape from MBUs. This is easier applicable to FGTMR, because in the case of CGTMR there is not any control on Fine Grains to make the best placement for redundant ones. This issue will be discussed in more detail in Chapter 6.

4.9 Discussion on fine grain fault tolerance

As mentioned, in order to be able to control and verify each node of the design, one must have a finer view to the FPGA structure. In this work it is called Fine Grain fault tolerance. The approach is based on one important feature of reconfigurable architectures, respectively SRAM-based FPGAs, namely their homogeneity [92].

Henceforth, the scale for mitigation will be limited to pairs of LUT and FF. By focusing on architectural design primitives it is possible to define localized redundant patterns for LUTs-FFs and apply these patterns to any design that is mapped to the FPGA. This approach needs to insert LUTs to the design which rule as voters. Then based on the TMR technique which is used, it can achieve more reliability. Every fine grain defined part, can deal with one SEU and much possible SETs as soon as not more than LUT in every TMR-LUT are the same time affected, which has a very low probability. In the big scale, the design will be available for a long time. SETs occur and are masked by local voters and SEUs are masked and are corrected in case of using feedback at their inputs [91].

However, in FGTMR a lot of fine grain voters are used which effect critical path strongly. The next step of this work will modify the voting structure of fine grains in the system, including evaluation on complex designs. The work on the integration of inherent voting for FPGA primitives, becoming an integral part of the framework is done as well. Still an open issue in the implementation are place and route constraints targeting to solve MBU effects.

However, although fine grain TMR is simple and applicable to every kind of design, using voters at the output of every fine grain can result in a big area overhead in the system. Because of this, the techniques which eliminate using voters may be interested in reliability approaches. Quadded redundancy which is based on inherent voting is the focus of this work and will be discussed in the next chapter.

5 Quadruplicate Force Decide Redundancy - A New Trend on Mitigation Techniques

In this chapter, the fundamentals of Quadruplicate Force Decide Redundancy (QFDR) are described. QFDR is a fault tolerance technique based on inherent voting. The proposed methodology is an extension of quadded logics which has been proposed for logic gates. In this work, the quadded logic methodology has been extended to boolean functions.

As mentioned in the previous section, [99] shows that, by partitioning the TMR in the design, the system will achieve higher levels of reliability. For example [99] applies the Markov model [83] to a system that is implemented by TMR to analyze the reliability. However, in high scales, several voting steps for a fine grain TMR results in a big overhead in area as described in the last chapter. This work is focusing on an alternative to fine grain TMR which does not need voters in the fine grain fault tolerance structure. QFDR is a methodology that uses the fundamentals of voting without requiring specific voter components in the output of fine grains.

5.1 Inherent Fault Tolerance in Logics

Inherent fault tolerance was originally defined for logic gates, as some gates the feature of error masking inherently exists within their boolean function.

In general, NOR, NAND, AND, and OR logic gates are the basic gates with this feature. In these logics, there always exists a dominant input which defines the output value. If there is one dominant input value in the inputs, other input lines will be masked due to the dominant input value which forces the output to a given value. For example, in a NOR gate with two inputs, as soon as one of the inputs has the value '1', the output is always '0' either the other input is '0' or '1'. This means as soon as a correct dominant input exists, the output is correct, independent of failures in the other lines.

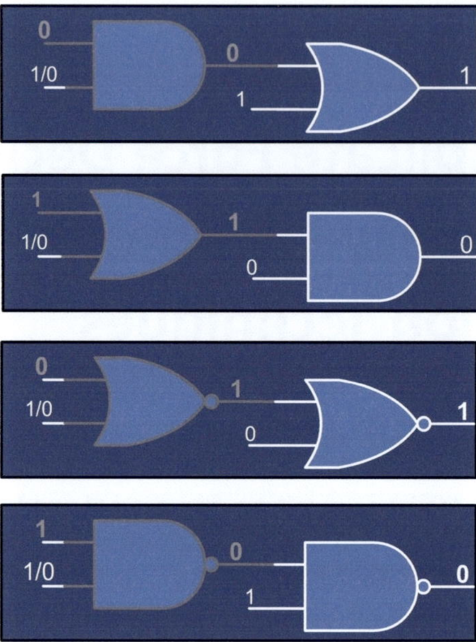

Figure 5.1: NAND, NOR, AND and OR logics in a combination which is used in quadded form. Dominant input values are having a deciding rule in the next level. Two inputs in every gate are duplicated.

If the dominant input value is faulty, the output will be forced to a faulty value while the other input lines have no control on it. In order to cope with this problem, two countermeasures can be taken to make the structure fault tolerant. Firstly, redundancy can be added to the design for saving the correct output value for example by duplicating the gates.

The second countermeasure refers to the structure of boolean function of the gate. If the faulty output value is a weak value and not the dominant value in the next logic stage (conversion of dominant value), it is automatically ignored in the next stage by using the duplicated correct and dominant output value. For example, in NOR logic, '1' is the dominant input value and as soon as it appears in one of the input lines, the output value will be '0' which is the conversion of '1'. This simple feature exists in NOR, NAND, and the combination of AND/OR logics. Figure 5.1 refers to this feature.

If two duplicated gate's outputs are used as inputs to the next stage gates, then the correct dominant output value defines the output of the new stage and in this way the fault is corrected.

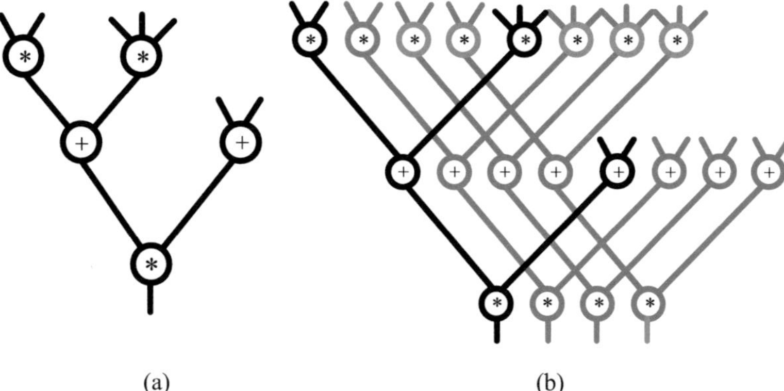

Figure 5.2: (a) a simple AND(*)/OR(+) logic Network (b)The basic quadruplication of gates

However, by duplicating the inputs and gates, there are still some scenarios in which the inherent fault tolerance will fail. When, a dominant input value influences both gates, the correct output value can not be saved anymore. Because the structure of masking and correcting is based on duplication, the number of gates can be evenly grown up. Due to this quadded logics are constructed which are based on quadruplicating the logic gates.

5.2 Quadded Logics

Quadded Logic (QL) is a redundant logical structure which is based on the previously described inherent fault tolerance feature in logic gates. It is proved to be tolerant against all single faults and most multiple ones [120][121]. To introduce the concept of quadded logic for gates, the easiest way is to build it up from some rather simple principles. The fundamentals as described in [121] include:

1. The logical circuitry is quadruplicated and the inputs are duplicated.

2. Any error can be corrected in the logic just downstream of the fault that caused it.

3. Correction is accomplished by good signals from the neighbors of the faulty unit on the same logic level.

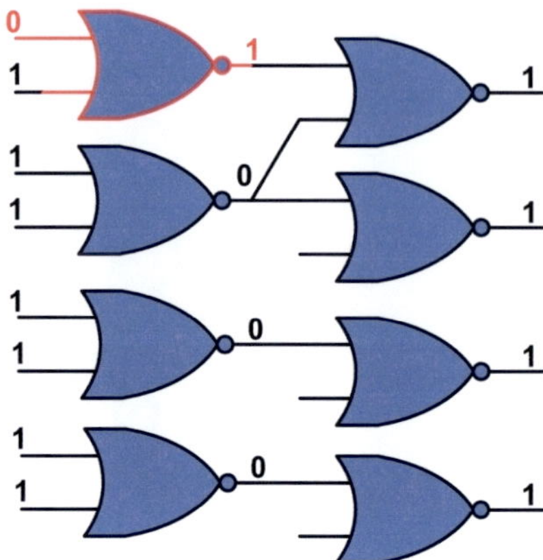

Figure 5.3: Duplicated ouput in the second stage defines the ouput value and masks the faulty input.

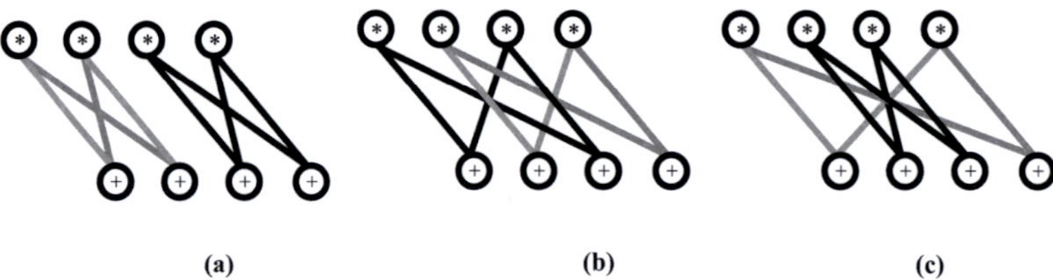

Figure 5.4: Different connection possibilities between two stages

4. In order to save the correct values, in every stage, two gates in addition to their inputs and outputs are independent from the other two gates in quadruplicated form.

These principles form the quadded logic fault tolerance structure. To get the nature of quadruplication, it is applied to a simple logical network. The original network in AND/OR gates is shown in Figure 5.2(a). The quadruplicated circuit is shown in Figure 5.2(b).

According to the first fundamental, the original network is quadruplicated. Four identical logic networks are fed with four identical inputs. If the inputs are correct and not faulty, they generate four identical outputs.

When an error occurs, every input line which has bit flipping can fulfill these two possibilities:

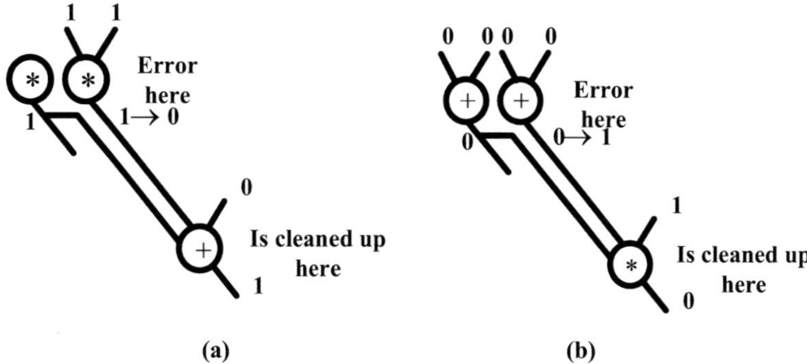

Figure 5.5: Mechanism of Error Masking in Quadded AND/OR Logic

1. The current value is not a dominant value as input of the gate. In this case, it will be ignored by the other dominant value and the error will be inherently masked and not propagated.

2. The current value is a dominant value as input of the gate. In this case, it will change the output to a wrong value. As described, this wrong value, due to the function of the gates which were listed, is surely not the dominant input value of the next stage. This means that, this wrong value will be ignored by the other correct input values from neighbors.

For example, as shown in Figure 5.3, the inputs are duplicated in a NOR gate and the NOR gate is also quadruplicated. When the original and not faulty value of inputs is '1', the output value will be always '0' even when one of them is '1' due to bit flipping. When the original value is '0', while bit flipping, the value will be changed to '0'. However, in the next stage, '0' will have a pair from one of the other quadruplicated gates. When all of other outputs from previous stages are '1' and just this one due to bit flipping is '0', then its pair will be '1' instead of '0' and forced the output to '0' and then the function of NOR helps ignoring the wrong value. This basic inherent fault tolerance can be defined also for NAND gates and AND/OR level gates in a similar manner, based on their special features.

In the AND/OR circuit in Figure 5.5 is shown how duplicating the inputs and the described features mask and correct an incorrect value in AND/OR levels.

The final quadded circuit is illustrated in Figure 5.6. Quadding method cleans failures up using the same feature. However, the key feature is that the wrong output value which is generated, is not the dominant input for the next stage and the neighbor correct output values are not the same as the wrong one. Due to this

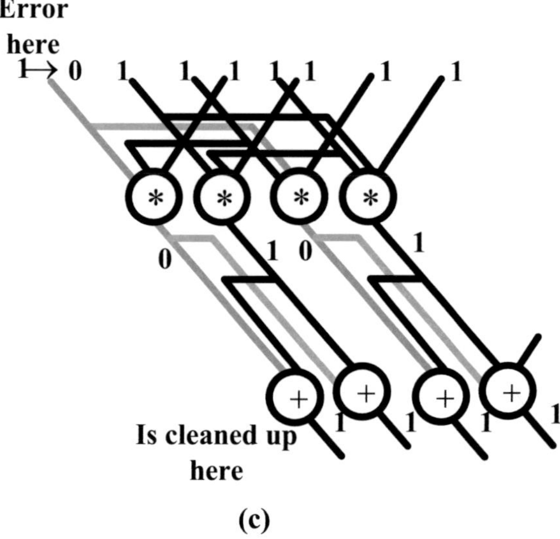

(c)

Figure 5.6: Final Quadded AND/OR Logic

fact, they can role as the dominant input value for the next stage [70][120][121].

It should be noticed that, in order to be operational, in quadded form, the output of each logic unit is joined at the next logical unit by the output of a neighbor. The pattern of the connection is systematic. Between each group of four logical units and the following group, there are four straight-through and four crossed connections. The error normally spreads into two channels before being corrected. If the connection structure remains similar till the end, it is possible that the error is propagated till the final stage. In this way of connection, the error will be blocked before propagating into the further stages. Different connection possibilities are shown for the simple circuit in Figure 5.4. More details on this way can be found in [120][121].

5.3 Quadded Force Decide Redundancy

In order to explore the idea of quadded logics in a general boolean function, some features have to be added to the general logic. The Quadded Force Decide Redundancy (QFDR) is a redundant logical structure which quadruplicates logical functions and defines two different *Force* and *Decide* rules for logic functions based on their level in the design and then connects them together using special connection patterns. While QL only allows simple gates like AND and

OR, that have to be interconnected in a certain manner, in QFDR, the behavior is generalized to all boolean functions.

The reason that quadded logic in its basic definition for gates does not directly work for boolean functions is that, the basic feature which makes quadded logics possible in NAND, NOR, and AND/OR logics, is their definition. As described, dominant inputs in these logics can define the output value. Quadded logic is constructed by using this feature to mask and correct errors. However this feature is missing for a boolean functions in this general definition. It is possible that for example a '0' is a dominant value in a certain input line but not the other input lines. Due to this fact, a more general description of quadruplicating is needed for boolean functions.

A general quadruple network is constructed using the following underlying assumptions: First, the logical functions must appear in quadruplicated form and all inputs must be duplicated. Every two out of four quadruplicated copies are the duplicated inputs of the next level. Second, errors are corrected in the logic just downstream of the fault that caused it and third, the connection is accomplished by correct signals from the neighbors (in duplicated inputs) of the faulty logic function.

This can be solved by the *Force/Decide* approach which is proposed in this work. It works as follows: Every input is duplicated and overall there are three possibilities for duplicated inputs. Either both are '1' or both are '0' or one of them is '0' and the other one is '1'. Any difference in duplicated inputs means that one of them has incorrect value.

In the first level the output is forced to a *one* in case of difference in two duplicated inputs. This means whenever the output is *zero* it is known that no error occurred. If it is *one* it might be correct or wrong.

This way the bit value contains implicitly some additional information which is used by the next level. The next level knows about the forced output value as the result of the previous level and in case of *one* the output value will not be assumed as correct value and respectively will not be selected as input value of the second level. Functions f and g are modified to satisfy the corresponding force and decide rules.

Figure 5.7 (a) shows function f containing two original inputs and in the quadruplicated form, four inputs with every two being equal (Figure 5.7 (b)). In this example duplicated inputs are (i_1 and i_3) and (i_2 and i_4). The modification of $f(i,j)$ and $g(k,l)$ is done for the force level function as follows (exemplarily shown for f_1):

$$k_1 = f_1(i_1, i_3, j_1, j_3) = \begin{cases} f(i_1, j_1) & (i_1 = i_3) \,\&\, (j_1 = j_3) \\ 1 & otherwise \end{cases}$$

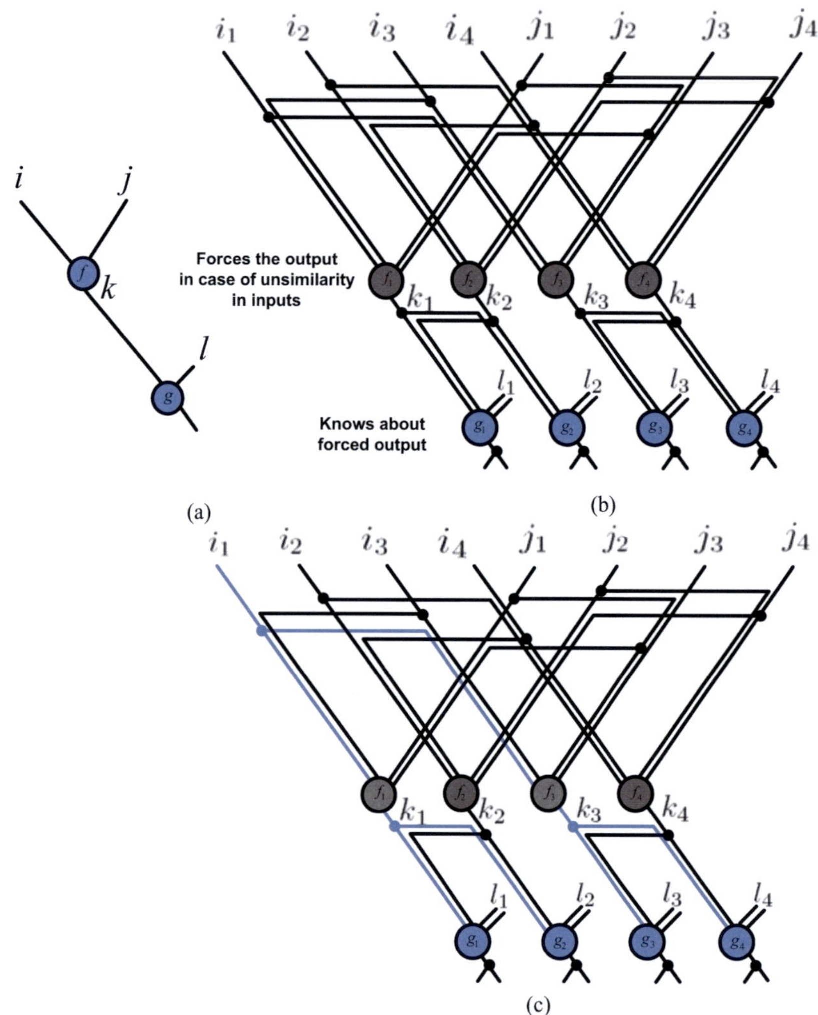

Figure 5.7: Quadruple Force Decide Redundancy for logic functions

and for decide level (exemplarily shown for g_1):

$$g_1(k_1, k_2, l_1, l_2) = \begin{cases} g(k_1, l_1) & (k_1 = k_2) \,\&\, (l_1 = l_2) \\ g(k_1, 0) & (k_1 = k_2) \,\&\, (l_1 \neq l_2) \\ g(0, l_1) & (k_1 \neq k_2) \,\&\, (l_1 = l_2) \\ g(0, 0) & (k_1 \neq k_2) \,\&\, (l_1 \neq l_2) \end{cases}$$

This method is applied to all f_is and g_is accordingly. The method is applicable to all n input functions, while the example depicts two input functions for simplicity.

Now assume a failure occurs at input i_1. It will not be equal to i_3 anymore. The output in this condition is set to a *zero* forced value. Two scenarios are possible. The first scenario is, that f_1 and f_3 are forced to the default value and it is equal to the correct output value which comes from f_2 and f_4.

Hence, the next level, which contains function g, will have four equal inputs from the previous level (f) and accidentally the forced value in this case was equal to the correct output value. The worst case scenario is shown in Figure 5.7(c) where the forced output value of f_1 and f_3 is not equal to the correct output of f_2 and f_4 . Because the next level including function g knows about the forced output value of the previous level, it will select the correct result in case of difference.

However, as for FGTMR described, all fine grain fault tolerant parts are small parts of the system and each of them can deal with one single failure without disturbance of the overall system. Due to this fact, the probability of multiple failure occurrences in two redundant fine grains is much less than the coarse grain one. Multiple distributed failures are harmful for CGTMR. They affect the fine grains as well, but the fine grains deal with the failures separately. In this way the fine grains redundancy protects the system against multiple distributed failures. Due to fine grain fault tolerance in both FGTMR and QFDR, the same model for FGTMR can be used for QFDR.

5.4 Realization of QFDR in FPGAs

FPGAs are homogeneous architectures which consist of LUTs/FFs in a fine grain view. As LUTs realize boolean functions, they can also be a part of QFDR in FPGAs. In order to apply QFDR to FPGAs, LUT's functionality is defined as boolean function and in addition with flipflops the LUTS are quadruplicated and connected in a manner which forms QFDR. This is the basic principle in applying QFDRs on FPGAs.

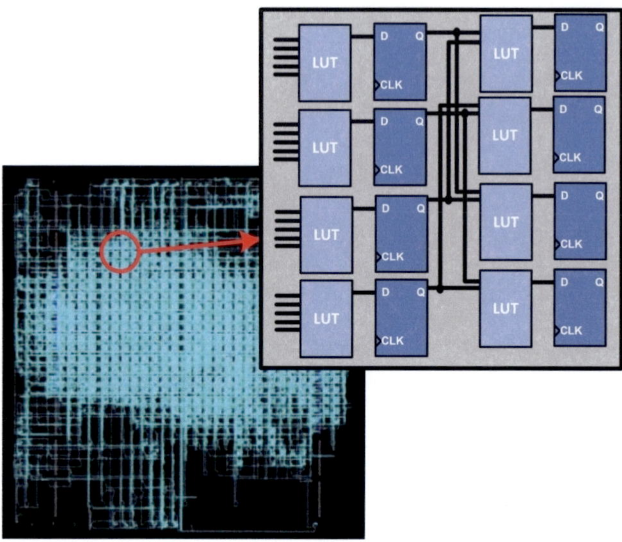

Figure 5.8: Every LUT and its corresponding flipflop is quadruplicated in QFDR in FPGAs

Before proposing QFDRs for FPGAs, it is needed to mention, even quadded logics in their traditional form are applicable on FPGAs. In this way, the fault tolerance targets the SETs which occur in logic and can be propagated to flipflops to cause SEUs. The combinational logics are modified to NAND, NOR, or AND/OR forms and then are quadded. However, this does not provide the expected result, as the synthesis tools automatically pack several logic layers into singular LUTs hence removing the inherent recovery feature. In addition one can expect, that such redundancy will be removed by the synthesis flow.

In order to realize force and decide rules in LUTs, the boolean function of each LUT needs to be modified. As described, input lines are duplicated in every function. When the LUT' role is a Forcer, every duplicated lines are compared and if they are different (one '0' and one '1') the corresponding input line will be forced to '1'. Two duplicated input lines A and B are replaced with: (A XOR B) OR f(A) and force the output to '1' as the value of input in case of difference.

If the LUT's role is a Decider, in case of difference, it will decide for '0'. Therefore, two duplicated input lines A and B can be replaced with: A AND B. This means that, if they are different, '0' will be set as input and if they are similar, one of them will be set as input.

This is the only modification which is needed in the function of LUTs. In addition, conncetions between LUTs are also modified in a manner, which completes the QFDR. Realization of QFDR on FPGAs and its limitations and requirements is described in detailed in the next chapter.

As Figure 5.8 illustrates, in a general form, LUT/FFs can be QFDRed. In this way, QFDR makes the combinational logic tolerant against SETs. Thus, no SET is possible to propagate in the logic and spread out to the next level of flipflop. When a SEU affects a flipflop directly, a permanent state change occurs in the flipflop. Due to QFDR, the next level will mask the wrong output from the faulty flipflop and avoid propagating it through the design. The faulty flipflop is corrected as soon as the fresh correct inputs are feeding it.

5.4.1 Flipflops without Enable

Flipflops in a design can be grouped into two different types. The first type comprises the flipflops which always change their state with each clock pulse. In the second type, there is an enable signal for the flipflops which activates or disactivates the loading of new values. This grouping is important due to SEUs. As previously described, SEUs can change the state of a flipflop. If a flipflop hold its state, it is possible that a SEU changes this state and is not further corrected by a good signal from the input of flipflop.

In the first type of flipflops, SEUs can change the state of flipflops. However this is ignorable because in the next clock cycle the correct value will be written again in the flipflop and corrects the state. Even when one of the flipflops is affected by SEUs, this will be duplicated in the next level of design and will be masked after an step of force/decide. Figure 5.9 illustrates the these kind of designs with QFDR.

5.4.2 Flipflops with Enable

In the cases of enable in flipflops, which are for example used in counters and state machines, simple QFDR may be unable to control the state of flipflops after a SEU changes the state of it and it remains as the state of flipflop till enable is further active.

The scenario is as follows: a flipflop is affected by an SEU. This flipflop doesn't use an input from the previous stage's combinational logic to be corrected, but rather, the state remains in the flipflop. In this case, it means that, in the case of affected flipflop, the state will be never corrected till the flipflop enable is activated and the flipflop is initiated to a correct value.

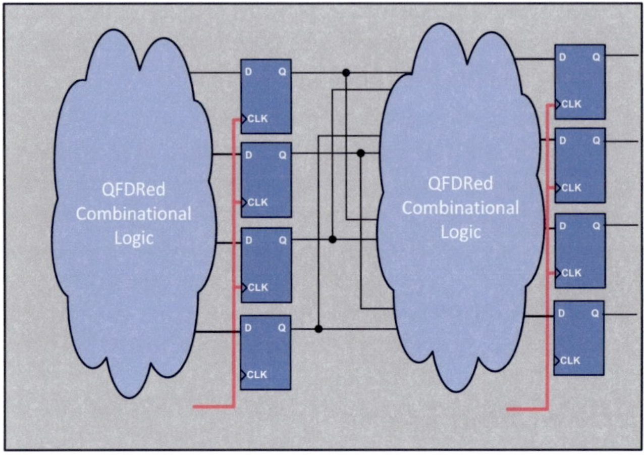

Figure 5.9: In the designs which their flipflops does not have any enable every, SEU which affects the flipflop will be rewritten in the next level.

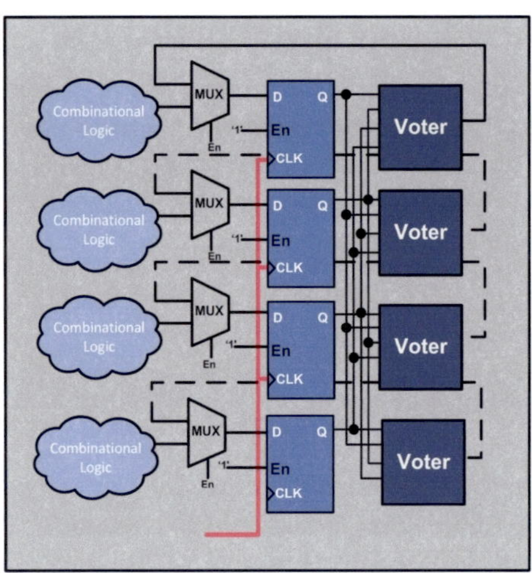

Figure 5.10: Flipflops which include enables to the corresponding combinational logic may be susceptible against SEUs even after QFDR. The wrong value will never be corrected if the flipflop be not again initialized when the enable is not active.

To cope with the SEU in flipflop with enable, one solution is to check the state of quadruplicated flipflops with a voter and as soon as a change occures in the state of one of them and its enable is not active, active the enable of it and rewrite the correct state from the other three flipflop copies. This can be done by using a multiplexer at the input of flipflop and voter as shown in Figure 5.10.

However, this problem is not limited to the QFDR. The other redundancy-based fault tolerance methods also have challenges to correct the flipflops with enables in case of SEUs change their states. Table 5.1 has measured the number of flipflops with feedbacks in different benchmarks. However, in the benchmarks which does not include state machines, about 15% of flipflops use feedback and the overhead of using voters at the end of them is ignorable in the whole design.

In QFDR the role of voter is actually distributed between the next level of logics. This means that, instead of inserting the voters in this case, one solution can be to use the output of Decider LUTs in the next level of combinational logic. However, this is more complicated and needs exact analysis of the design to avoid some wrong scenarios. As a result it does not have the overhead of using voters in a QFDRed design.

Table 5.1: Amount of flipflops with Enable in different designs

Design	All Num. of FFs	FFs with Enable	Percentage(%)
Cordic	2159	265	12
MotorCtrl	543	26	4.7
uProc.	773	44	11.5
FuzzyCore	1259	145	4.1
PicoBlaze	241	10	4.1
Counter	15	2	13.4

5.4.3 One Stage Combinational Logic

In some special cases, combinational logic includes not enough stages to be able to form Force/Decide structures. In this cases, some extra logics can be added to the combinational logic which function as a Decider.

Figure 5.11 shows this condition. The combinational logic is modified by inserting a Decider Stage. The next flipflops will use the output of the Decider. The output can be used also back in the combinational logic in cases of feedback for example to balance feedbacks. So FGTMR can also be used in this situations.

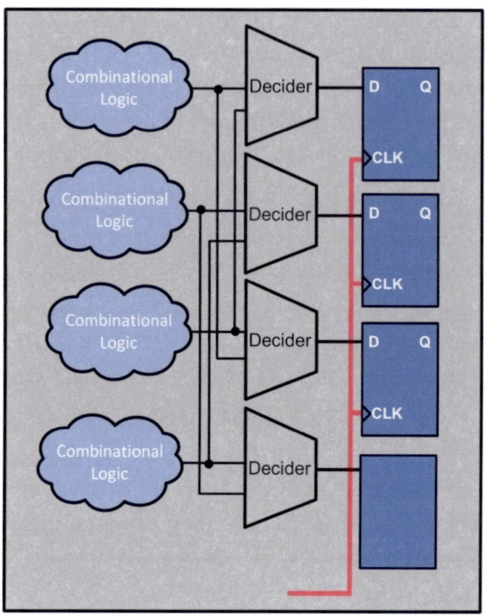

Figure 5.11: Inserting a Decider Level in case of small combinational logic

5.5 Reliability of QFDR

The exact analysis of reliability of quadded logic method has been regarded as extremely difficult and costly [14][119][71]. The reliability of a network of QFDR is the probability that it functions correctly. One idea is to calculate it as a function of the reliability (or the failure probability) of a fine grain component (LUT/FF) with certain assumptions. If a component fails so that it gives an intermittent or transient error at its output, this will be correctly immediately unless its neighbor gives a wrong output at the same time. The probability of error occurring for two neighbor components in the same time period is very small. Furthermore, their permanent effect on the system is realized when SEUs occur in flipflops and are permanent. Therefore, the SET effects in this analyze is reduced to the SEU results on flipflops which are permanent failures.

QFDR is designed to correct all single incorrect bit flipping and many multiple ones in a network of LUT/FFs if the proper interconnections are made, i.e. if the pattern at the output of a quadruplicate component is different from the interconnection patterns of any of the quadruplicate series at its input as already mentioned.

In this reliability analysis, the quadruplicated form of a LUT/FF pair is called a quad and is shown as Q. Because of the symmetry of the components in a

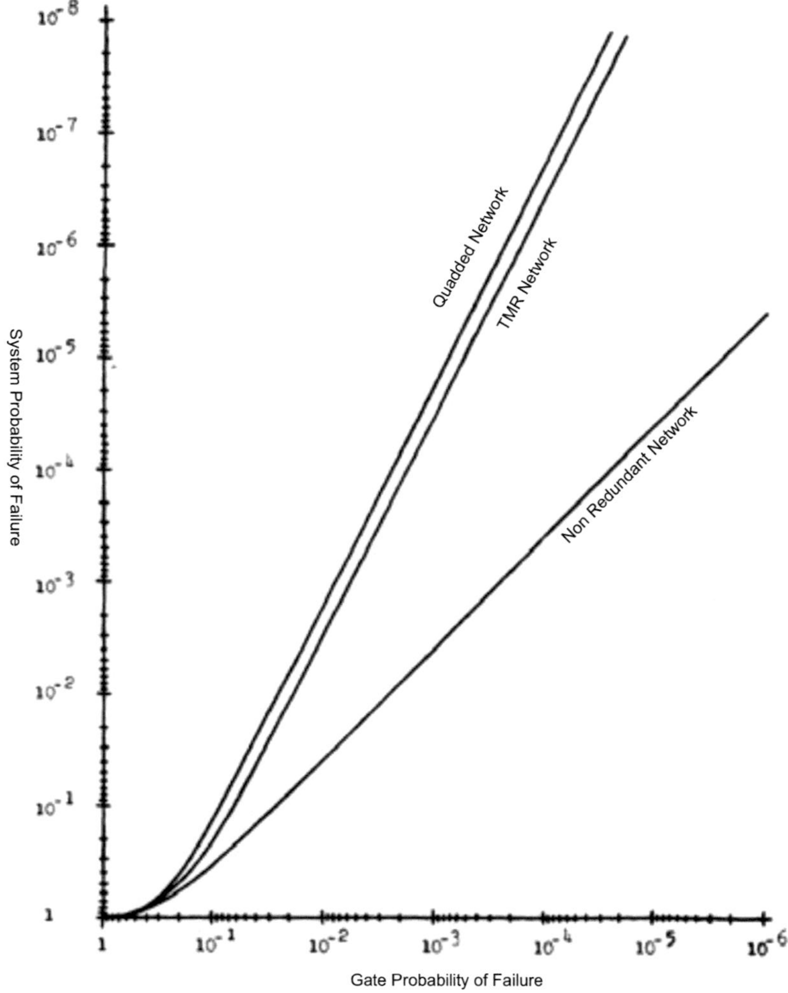

Figure 5.12: Reliability of a quadded design vs. the probability of logic failure. It is compared to the corresponding TMR network and non-redundant circuit [14].

quad, it is not needed to know which one of the quadruplicated components are faulty. Therefore a critical fault pattern which can make system failures can be concisely denoted by $\left\{ Q_1^{j1}, Q_2^{j2}, ..., Q_n^{jn} \right\}$ where $j_i \in \{1,2,3,4\}$. This model of presentation is already used to analyze the reliability of quadded logics [14]. Q_i^{ji} means that quad Q_i has j_i critical faults. Basics of quadded logic reliability analysis and QFDR reliability analysis are similar because the QFDR is derived from quadded logics and changes the function of LUT to be passable to the quadded logic features. Due to this fact, their reliability analysis have the same steps.

A quad Q_i is said to feed another quad Q_j if there are lines connecting the outputs of Q_i to the inputs of Q_j. If two quads Q_i and Q_j feed the same quad, this relationship may be denoted concisely by $Q_i F Q_j$. Two quads are said to be dependent if $Q_i F Q_j$ or if there exists a sequence of quads in the set. As in the definition of QFDR is described, a fault in a QFDR network is tolerable if all possible output failures in a quad are not the dominant input value for the quad which is fed by it. In this way, the reliability of QFDR network is defined as the probability that just the tolerable faults occur in the QFDR network. Therefore, in order to analyze the reliability of QFDR, it is needed to find all the tolerable fault patterns of the QFDR network.

The results are formalized in the lemmas below. The proofs are found in [15].

Lemma1: Given a set of dependent quads $\{Q_1, Q_2, ..., Q_n\}$, the number of tolerable fault patterns of the type $\left\{ Q_1^{j1}, Q_2^{j2}, ..., Q_n^{jn} \right\}$ is equal to

$$2 . \prod 2/j_i j_i \in \{1,2\}$$

Lemma 2: Subcritical faults are those that are not dominant and ignored by the next level in the existence of another dominant input value. Given a fault pattern $\left\{ Q_1^{j1}, Q_2^{j2}, ..., Q_n^{jn} \right\}$, the number of ways in which a quad Q_i can have one or two subcritical faults, which are tolerable, denoted by l_i and m_i, respectively, is as follows. If $Q_i \in \{Q_1, Q_2, ..., Q_n\}$, then if Qi is fed by any of the quads in the set,

$$l_i = 2/j_i$$
$$m_i = 2/j_i - 1$$

In the case that Q_i is not fed by any of the quads in the set,

$$l_i = 2/j_i + 1$$
$$m_i = 4/j_i - 2$$

In the case that $Q_i \notin \{Q_1, Q_2, ..., Q_n\}$ but is fed by one of them,

$$l_i = 2$$
$$m_i = 1$$

and finally if Q_i is not fed by any of the quads in the set,

$$l_i = m_i = 4$$

for a critical fault pattern $\left\{Q_1^{j1}, Q_2^{j2}, ..., Q_n^{jn}\right\}$ two vectors can be defined which give the number of ways in which each quad can have one or two subcritical faults respectively [14]. As [14] describes, when there is a network of n quads, the topology of the network may be represented by the structure of matrix S of this network. An $n \times n$ matrix includes n quads which feed n other ones:

$$S\,(i, j) = 1, \text{ if Quad } Q_i \text{ feeds } Q_j$$

$$S\,(i, j) = 0, \text{ otherwise}$$

This matrix makes it easy to partition the network to the independent quads. The set of tolerable fault patterns of the network is given by the fault matrix F which is defined to be a *(2n + 1) X (2n + 1)* matrix such that

F(i,j) = number of ways in which exactly i critical and j subcritical tolerable faults may occur in the network. The theorems below show how may be obtained for a network [14][15]. Using the proofs of [15], these three theorems are used to estimate the reliability of a QFDR network.

Theorem 1: For a critical fault pattern $\left\{Q_1^{j1}, Q_2^{j2}, ..., Q_n^{jn}\right\}$ where the set $\{Q_1, Q_2, ..., Q_n\}$ can be partitioned into *m* independent classes, the number of ways in which the tolerable critical faults can occur is given by:

$$G_c = \Pi_{allmclassesk} \left[2 \cdot \Pi_{j_i \in \{classk\}} 2/j_i\right]$$

Theorem 2: Given a critical fault pattern with sets L and M, the number of ways which exactly k tolerable subcritical faults can occur is given by $G_c(k)$

Theorem 3: The entries of the fault matrix are given by

$$F(i, j) = \Sigma_{allcriticalfaultpatterns} G_c(j)$$

and finally, the reliability of a QFDR network is given by:

$$R_Q = \sum_{0 \leq i \leq 2n} \sum_{0 \leq j \leq 2n} F(i,j) \cdot p_1^i \cdot p_0^j \cdot R^{4n-(i+j)}$$

where p_1 probability that a component fails as 0 is flipped to 1

p_0 probability that a component fails as 1 is flipped to 0

R probability that the component does not fail.

The probability of failure of the quadded network and its comparison to TMR and non-redundant one is depicted in Figure 5.12. However, as described, because QFDR is derived from quadded logics, the probability assumptions and network presentations remains the same for QFDR networks and due to this fact, Figure 5.12 still is referable when analyzing QFDR reliability.

In addition it is needed to prove the concept of QFDR by applying it to some benchmarks. Firstly it must be integrated in our synthesis tool and secondly be proved by experiments. In the next chapter, this issue is discussed in detail.

6 Integration of QFDR in Synthesis Flow

In this chapter, the approach which is used to integrate QFDR in the synthesis flow of FPGAs is discussed. The implementation is based on but not limited to Xilinx Virtex5 FPGAs.

6.1 Synthesis Flow

The traditional approach to implement logic designs on FPGAs is started with a description of design in any kind of hardware description language (HDL) and then goes through series of steps to generate an output bitstream. This bitstream is further used to configure the FPGA with the design. This steps are called Synthesis flow and is generally the same for different kind of FPGAs families and their corresponding tools.

The first step is to take a logic circuit described in an HDL and convert it into a graph of logic gates. This graph is then optimized using various algorithms in a step called Logic Synthesis. Following Logic Synthesis, the Technology Mapping

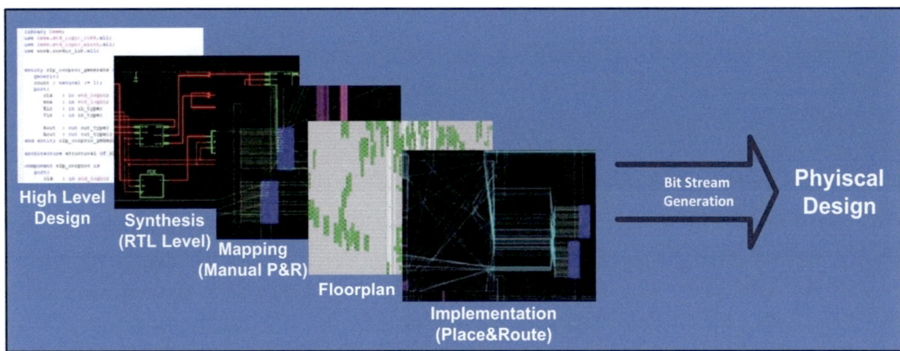

Figure 6.1: Synthesis Flow in Xilinx ISE and a lot of other Synthesis Tools

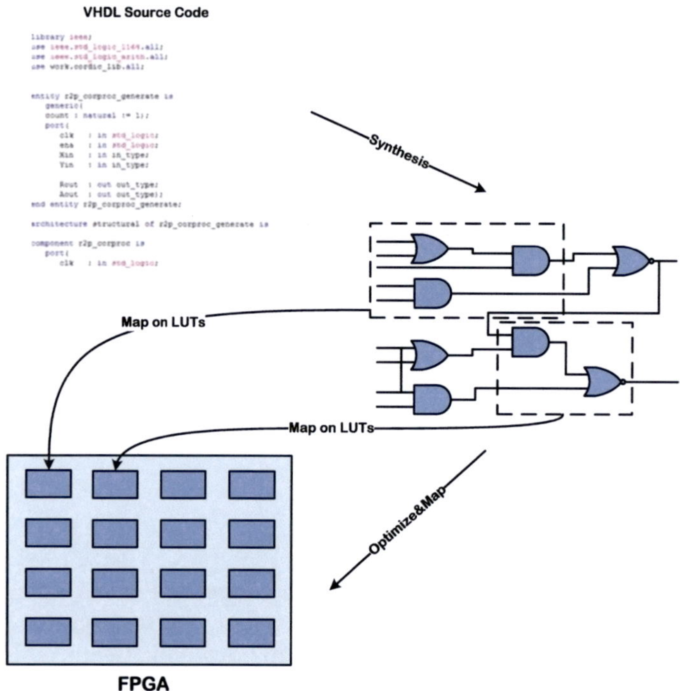

Figure 6.2: Structure of Synthesizing a design and mapping it on FPGAs

step takes an optimized graph of gates and represents it as a graph of resources available on an FPGA. These resources are usually Lookup Tables, Memory and Input/Output pins.

General logic functions are usually implemented as Lookup Tables, data storage modules are assigned to memory blocks and external connections are facilitated by I/O pins. A graph of hardware resources can then be placed on an FPGA, and the links between each node of the graph are realized by routing the physical connections between logic components using a programmable routing network. This step is called Placement and Routing.

Finally, the resulting implementation is analyzed by using a timing analyzer. A Timing Analyzer is a tool that computes the worst case delay information and determines the maximum clock frequency at which the circuit can operate. Results from the Timing Analyzer complete the timing analysis step, which is then followed by generation of a programming bit stream [30]. The above steps are depicted in Figure 6.1.

In order to apply QFDR to a design, it must be integrated in the synthesis flow. This means that, the suitable level of circuit description must be used as the input of QFDR. The QFDRed form will be further used as the input of the next level.

In theory, automatic insertion of QFDR can be done in different levels of synthesis flow. However, in the behavioral description of the design, homogeneous structures do not exist in every step.

In the synthesis result, the behavioral description is decomposed to smaller and more homogeneous logic gates. Although the netlist description in this level is optimized, more optimizations is possible during the technology mapping, in order to fit the design a given size FPGA resources. If QFDR be inserted to the synthesized netlist, it is possible that it is reduced due to the optimization during technology mapping. In Figure 6.2 the structure of synthesis and mapping is shown in more details. However, the result of synthesis is not homogeneous and the netlist parts sometime will be decomposed or recomposed after mapping. Due to these facts, circuit may not be a suitable input for applying QFDR directly after high-level synthesis.

The circuit description after technology mapping, is a netlist of FPGA resources which is a homogeneous description. It is a suitable input for QFDR. In modern Xilinx FPGAs, placement is done during the technology mapping. This is a disadvantage for QFDR, because the new copies which are added due to QFDR can not be automatically placed on the suitable resources by the original tooling.

Another possibility for applying QFDR is the routed description of the design. If QFDR copies are inserted after routing, the routing of inserted copies must be done manually. This is a time consuming process and may result in an unoptimized routing.

The last level of a synthesis flow is the bit stream generation. In theory QFDR can also be inserted after bit stream generation. However, the structure of bit stream is not clear and homogeneous structures can not be detected inside it.

In conclusion, QFDR is inserted to the synthesis flow after technology mapping in this work. This chapter focuses on the mechanism and structure of integrating QFDR to the standard synthesis tools.

6.2 Targeted FPGA family

This work targets Xilinx FPGA families and uses the corresponding ISE synthesis tool [64] to insert automatic redundancy. In this section, an overview of the Virtex-5 families is briefly described and then it is integrated to the radiation tolerant architecture by integrating QFDR on design synthesis flow.

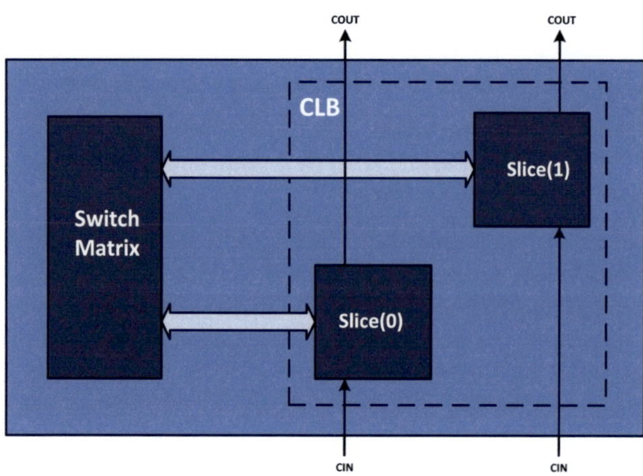

Figure 6.3: Arrangement of Slice with CLBs and carry chains [63].

6.2.1 Xilinx Virtex-5 Architecture

The Virtex-5 is an FPGA Virtex family that consists of two major components: Configurable Logic Blocks and Digital Signal Processing element.

6.2.1.1 Configurable Logic Blocks - CLB

The CLBs are the main logic resources for implementing sequential as well as combinational circuits. There are switch matrices at the neighborhood of CLBs. Switch matrices are used for routing. Every CLB contains two slices. Each slice has a specified position which is labeled with X and Y. Every Slice contains four Look Up Tables that are followed by flipflops. As already discussed, these are the basic elements which are used in QFDR on FPGAs. Figure 6.3 shows the CLB form in Virtex5 FPGAs and their connection to switch box and Figure 6.4 shows the X and Y positioning of slices and their connections together.

In addition to this, some slices support two additional functions: storing data using distributed RAM and shifting data with 32 bit registers. Slices that support these additional functions are called SliceM; others are called SliceL. An overview of two slice types is shown in Figures 6.5.

- **Look Up Tables** Look Up Tables are the essential elements in every slice to implement functions. As mentioned, there are four LUTs in every slice. Every LUT can be implemented as six-input function. There are six inde-

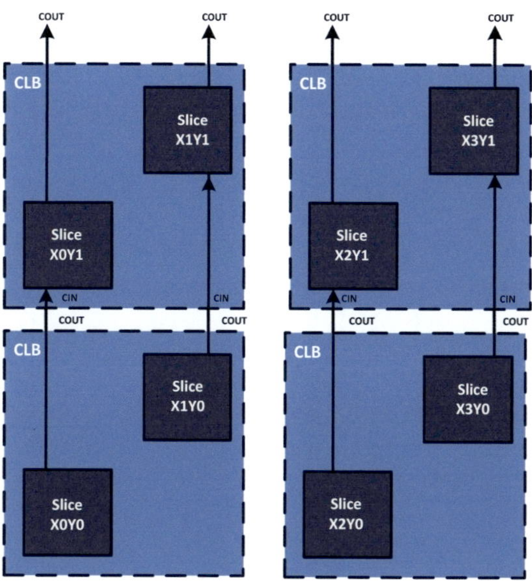

Figure 6.4: Row and Column Relationship between CLBs and Slices [63]

pendent inputs (A inputs - A1 to A6) and two independent outputs (O5 and O6) for each of the four LUTs in a slice (A, B, C, and D). The LUT can be implemented as any arbitrarily defined six-input Boolean function. Each LUT can also be implemented as two arbitrarily defined five-input boolean functions, as long as these two functions share common inputs. Only the O6 output of the LUT is used when a six-input function is implemented. Both O5 and O6 van be used for each of the five-input function which is implemented [63].

In addition to the basic LUTs, slices contain three multiplexer. The multiplexers combine up to four LUTs in a slice and provide seven or eight input LUTs.

- **Distributed RAMs** In sliceM, multiple LUTs can be combined in various ways to store an amount of data. The LUTs in SliceM can be implemented as a distributed RAM element. Based on the number of LUTs which are occupied, different configurations for RAMs (with different spaces) can be constructed.

- **Shift Registers** SliceM LUTs can also be configured as a 32-bit shift register. By combining the LUTs in a SliceM, it is even possible 128-bit shift register. Bigger ones can be constructed by combing a couple of slices.

- **Digital Signal Processor - DSP** In addition to the CLBs, Virtex-5 FPGAs are equipped with DSPs. In Virtex-5 DSP blocks (DSP48E), the 18x18 multipliers have been replaced with asymmetrical ones (18x25 bits signed). This reduces the DSP cost of floating-point single precision (24-bit significant) from 4 to 2.

6.3 Integration of QFDR on Virtex-5 FPGAs

In the previous chapter, the fundamentals of QFDR were described. In QFDR approach, every element of a design must be quadruplicated. In addition the inputs must be duplicated and the function of element must be modified depending of its role in QFDR.

Designs on Virtex-5 FPGAs are mapped on LUT and flipflop pairs, depending on the mapping strategy which is used. When area optimization is targeted, normally the design will be mapped in the least number of LUTs as possible. This means that, as much as possible number of inputs are used in LUTs to realize the function and optimize the design.

One of the main challenges in integrating of QFDR is here. LUTs in Virtex5 FPGAs contain six inputs. In order to be able to duplicate the number of inputs in QFDR, every function is allowed to use maximum three of six inputs in every LUT. For example, if a LUT in original design, realizes a function with three inputs and then the number of inputs are duplicated, all six inputs are used to realize QFDR.

Therefore, all functions must be decomposed to three or less input functions in order to realize QFDR.

6.4 LUTs Decomposition

6.4.1 Problem Formulation

Given an architecture A as the input of QFDR, which is used to implement different function of up to K variables, a function $f(X)$ is regarded as a wide function, if $|X| > K$. For Virtex5 FPGAs, K is equal to 3.

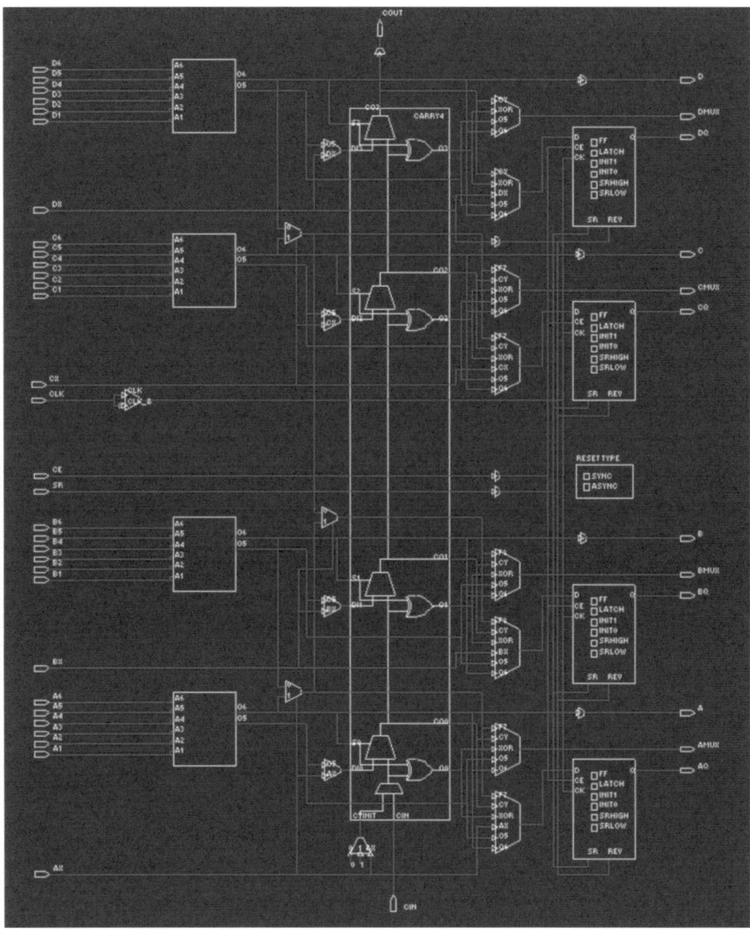

Figure 6.5: Diagram of SliceL. SliceM has the same structure as SliceL in addition to the feature that LUTs can be implemented as distributed RAMs in SliceMs.

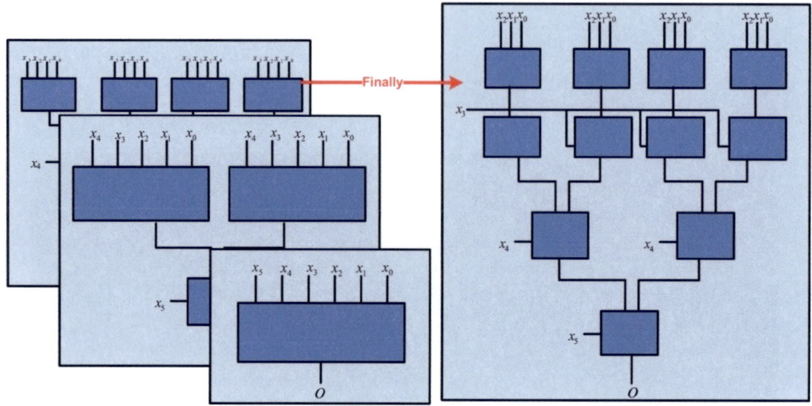

Figure 6.6: A 6-LUT can be divided to 3-LUTs using shannon's expansion theory

6.4.2 Shannon Decomposition

Shannons decomposition technique is applied to the LUTs in FPGA with more than three inputs. Afterwards, every LUT in FPGA has three or less inputs and the new design is ready to be modified by QFDR.

The approach is based on the well-known Shannon's theorem [11], which is briefly described here. Using Shannon's theorem, any Boolean function, g, of n variables, $g = f(x_1, x_2, ..., x_i, ...x_n)$, can be decomposed with respect to one of its variables, x_i, and written as the logical OR of two sub functions:

$$g = x_i.f(x_1, x_2, ..., 1, ..., x_n) + \bar{x}_i.f(x_1, x_2, ..., 1, ..., x_n) \qquad (6.1)$$

where $f(x_1, x_2, ..., 1, ...x_n)$ is called the 1-cofactor of f with respect to variable x_i and $g = f(x_1, x_2, ..., 0, ...x_n)$ is called the 0-cofactor.

6.4.3 Area Overhead of decomposition

Figure 6.6 shows the LUTs which are decomposed to 3-LUTs. In the first step of decomposition, a huge number of LUTs are needed to realize the decomposition as in Figure can be seen.

However, there are some methods which can significantly reduce this area over-head. During the decomposition and due to the Shannon approach, several cases

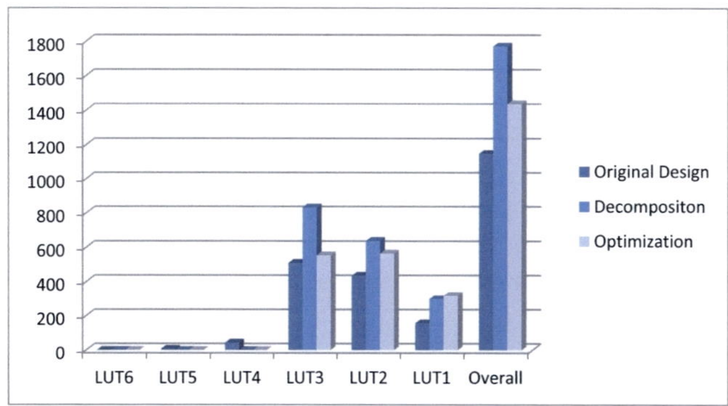

Figure 6.7: Original Cordic Design Overhead is compared to after decompositon of 6-, 5- and 4-LUTs and to the result of optimizations.

of functions will be ANDed to zero or will be ORed with one and this results to a lot of function eliminations. A Cordic design is used as a test case. It includes 7 numbers of 5-LUT and 44 numbers of 4-LUT in its original design. As in Figure 6.7 is shown, the number of LUT usage is enormously increased in order to decompose 4-LUTs, 5-LUTs, and 6-LUTs. After optimization, the area overhead can be extremely reduced.

However, the optimizations which are done, are locally for each LUT and cover the local functions only. If a multi step optimization strategy is employed and a global view to the optimization problem, like what is done during synthesis flow, the design can be optimally implemented using 3-LUTs with much less overhead.

In order to proof this issue, another experiment is done. The benchmarks are synthesized in Spartan3 using XC3S50 [66] and then are used for technology mapping for Virtex5 devices.

They key improvement is that, the size of LUTs in Spartan 3 family is maximum 4-LUT. If the NGC format of Spartan 3 be further used in Virtex 5 technology mapping, less 6-LUTs will be generated and this mean less area overhead in the whole design. In Table 6.1 a comparison between using Virtex 5 in the whole synthesis flow and using Spartan 3 in synthesis is shown. Results show a huge area optimization in case of using Spartan 3 in synthesis flow. Results in case of big benchmarks for example Fuzzy Logic Core shows more than 50% area optimization after using Spartan 3 to generate synthesied design.

This means that, if it is possible to optimize globally, like what is done automatically in synthesis for Spartan3, the area overhead of decomposition should not be a huge matter anymore. As in Table6.1 can be seen, the area overhead which

Table 6.1: Area overhead in the original design, after modifications which is needed before applying QFDR, and after applying QFDR. Comparisons have done for Virtex5 FPGA in the whole synthesis flow and in a second case in order to optimize number of LUTs with more than 3 input, synthesis is done for Spartan 3 and technology map is further continued for Virtex 5 FPGA.

| | Cordic | | MotorCtrl | | uProc. | | Fuzzy_Core | | PicoBlaze | | Counter | |
	S3	V5	S3	V5	S3	V5	S3	V5	S3	V5	S3	V5
Original	1162	1162	314	274	383	387	614	630	138	138	8	8
Modified	1572	1572	672	1084	1035	1657	1464	5114	284	284	22	22
QFDRed	6288	6288	2688	4336	4140	6628	5856	20456	1136	1136	88	88

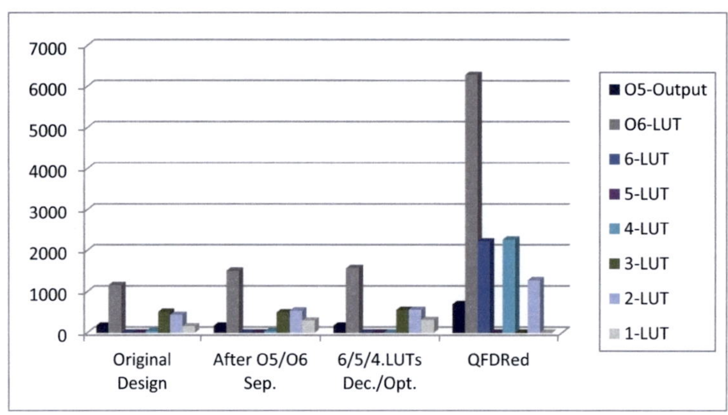

Figure 6.8: Area Overhead after decomposition and after QFDR in term of LUT types.

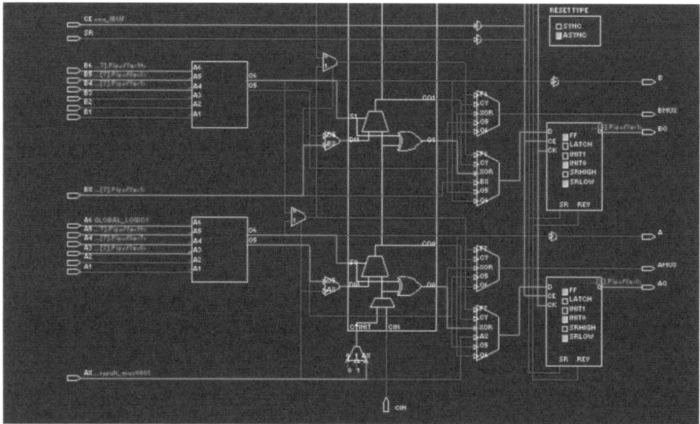

Figure 6.9: On the downer LUT both 5 and 6 outputs are used

is the result of modifications on LUTs is much more than the overhead of QFDR.

However, [11] has shown, if the k-LUT is divided to several n-LUTs, where n<k, the number of SRAM cells which is used, is a sight better than the case which k-LUT is used for mapping.

6.4.4 LUTs with two functions

As described, it is possible to implement two independent function in one LUT as long as these two functions share common inputs. However, in order to apply QFDR, every LUT must realize one function because QFDR modifies the functions in LUTs. Due to this, if one LUT has used both O5 and O6 independent outputs, it must be split to two LUTs. The separation methodology is shown in Figure 6.10.

However, due to the design limitations, it is needed to connect the O5 outputs to VDDs in the Area Overhead results in Figure 6.8, after separating O5 and O6, there are still O5 outputs there which refers to this issue.

6.5 LUT Function Modification

In order to quadruplicate a LUT, firstly, inputs must be duplicated. After decomposing a LUT, maximum three inputs are used which are A1, A2, and A3. A1, A2, and A3 are duplicated respectively in A4, A5, and A6.

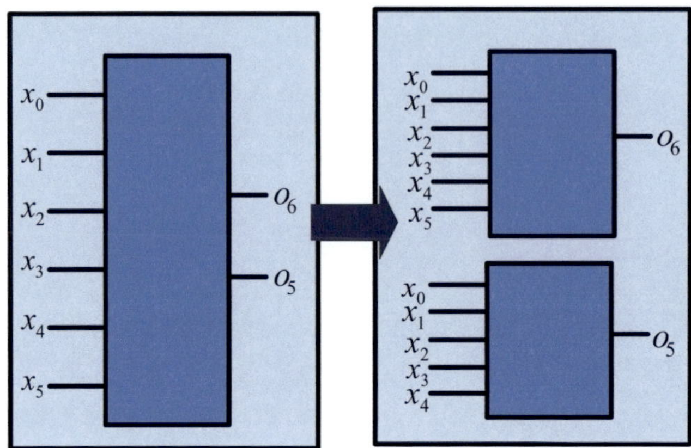

Figure 6.10: LUTs which use both O5 and O6 are separated as illustrated

Depending on the stage of LUT in design, it is role as Forcer or Decider. When a LUT is a Forcer, it means that, when its duplicated inputs have different values, or in other words, one of the inputs is faulty, the corresponding output will be forced to '1'. In this way, two duplicated inputs are XORed together and are replaced on the original input.

For example, if a LUT is fullfill the following function,

$$O6=(((A1*A2)+(A1*A3)))$$

The forced LUT will be modified to:

$$O6= (((A1*A2)+(A1*A3)))+((A1@A4)+(A2@A5)+(A3@A6))$$

where *, +, @, and   are respectively the AND, OR, XOR, and NOT symbols in NCD description of LUTs.

If no error exist, duplicated inputs have the same value. If an error occurs, the duplicated inputs will not be similar any more. In this case, if they are the inputs of a Forcer LUT, they both will be forced to '1', based on the fundamentals of QFDR.

Otherwise, they are the inputs of a Decider LUT, and '0' is decided to be the common value of the inputs. In other words, if A1 and A4 which are duplicated inputs have different values '0' and '1', '0' will be decided as the correct value. Hence, A1 will be replaced to (A1*A4) where * is the AND symbol.

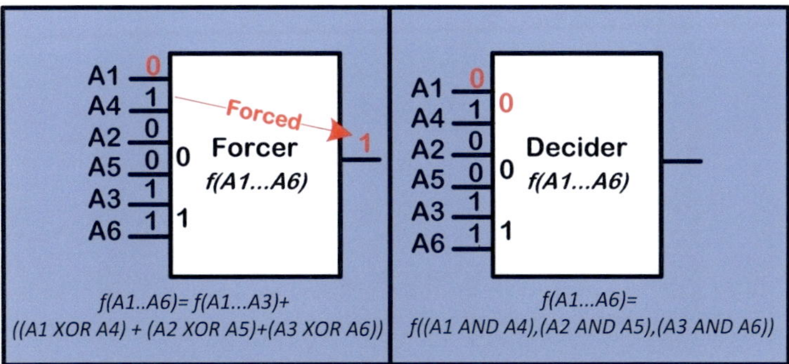

Figure 6.11: Forcer and Decider implementation in a LUT

Therefore, the same function in the decide rule will be modified as the following:

$$O6=(((A1*A4)*(A2*A5))+((A1*A4)*(A3*A6)))).$$

In both Forcer and Decider LUTs, a comparison between two duplicated inputs is done and based on the result of comparison, a decision is made. This comparator is the basic structure of Forcer and Decider LUTs. Figure 6.11 shows this for Forcer and Decider LUTs.

In the first version of QFDR on Virtex-5 FPGAs, the forced and decided roles are applied on LUTs in SlicesL and SliceM. However, the other elements, like Memories, shift registers, and DSPs, are quadruplicated to feed the inputs to the corresponding LUTs. This is a correct modification, because if one element such as DSP is affected, the forced/decide methodology which is used in LUTs can mask it and it will not be propagated through the circuit. In the input, they have normally voters which decide the correct value in four prepared outputs from the previous step.

6.6 Flipflops with Enable

LUTs are quadruplicated as a pair of LUT/FFs. However, flipflops which use enables, are those which need special voting wire as in the previous chapter described. Here if a SEU changes the state of flipflop, it will not be corrected till the enable is active. Here, an extra hardware is inserted to the input of flipflop (Chapter 5, Figure 5.10)

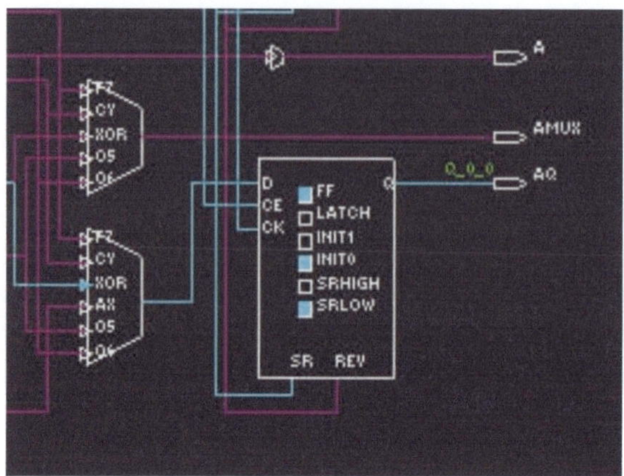

Figure 6.12: CE is active enable in FF in a part of SliceL.

6.7 Rapid Prototyping Platform

As already described, the logic optimization of the mapping step removes at least parts of the redundancy due to logic optimization. This was the main reason to apply QFDR to the mapped (and for new Xilinx architectures also placed) design. The result of map, which is a netlist in *.ncd* format is taken and converted into ASCII *.xdl* [22]. In the next step the redundancy is automatically added to the *.xdl* design description. Then the resulting netlist is converted back into *.ncd*. This can be used within ISE again for Place and Route and finally bitstream generation.

6.7.1 Xilinx Design Language

Xilinx Design Language (XDL) is a presentation of design on Xilinx FPGAs in general and an alternative (or a conversion) to the native netlist format called NCD. NCD is the result of synthesis, map, and place and route on FPGA which gives a netlist format of resources. This netlist can be presented in text form as XDL. An XDL file represent a design using 4 different statements [22]. The design statement, the module statement, the instance statement, and the net statement as shown in Figure 6.13.

6.7.1.1 Design Statement

To represent the design in XDL format a Design statement is used and is started with the keyword *Design*. It contains the design name and the name of targeted FPGA. Figure 6.13 shows an abbreviated XDL file with a design with the name *adder4bit* targeted for the FPGA part *xc4vsx35ff668-10* [50].

The design statement includes instance of primitives which are represented in instance statement. These primitives are configured to provide logical functionality. It also contains nets to logically interconnect the instanced primitives together. A design may contain one or more module statements. Every module statement includes instances of primitives and nets to connect it to the other modules. Nets in the design also interconnect the Modules. The design Statement, is modified in the number of instances in a module, also the number of inputs and outputs which are quadruplicated in QFDR.

6.7.1.2 Instance Statement

In the XDL file the Instance statement starts with the keyword *inst* and it instances a FPGA primitive. The instance statement contains a field for the instance name, which has to be unique in a design. The instance statement also contains a field for the primitive type, which can be a SliceL, SliceM, DSP48, or RAMB16. Figure 6.13 shows an instance with the name *adder1* of primitive type SliceL.

The instance statement also contains a configuration string, which starts with the keyword *cfg* to describe the primitive. The configuration string contains several attribute strings with the format *attribute name:logical name:value* and each of these attributes configures a part of the primitive. The attribute name represents a part of the primitive that is being configured and generally it is the name of a physical gate in the primitive.

From Figure 6.13 the instanced primitive's configuration string contains an attribute string *DXMUX::#OFF*, which indicates that the *DXMUX* attribute representing *DXMUX* gate is switched off and the logical name is left blank. Primitives are configured with the help of different attribute strings, which together form the primitive configuration string. In this way, logic blocks of different types like adders and counters can be made out of instanced primitives with each of them configured using different configuration strings.

The instance statement also contains a field for the placement information if the primitive is placed in one of the Tiles that can contain that primitive type. In Figure 6.13, the primitive is placed in location (primitive site) *SLICE X1Y74* and that represents the first SliceL from the left of the FPGA and 74th SliceL from the bottom of the FPGA. In the same figure, the same location is also represented in terms of a Tile location *CLB X1Y37*, which indicates that the Tile is a CLB and the first one from the left of the FPGA and 37th from the bottom of the FPGA. Placement modifications can also be done in instances. This issue which is used in MBU fault tolerance has been later described in more details.

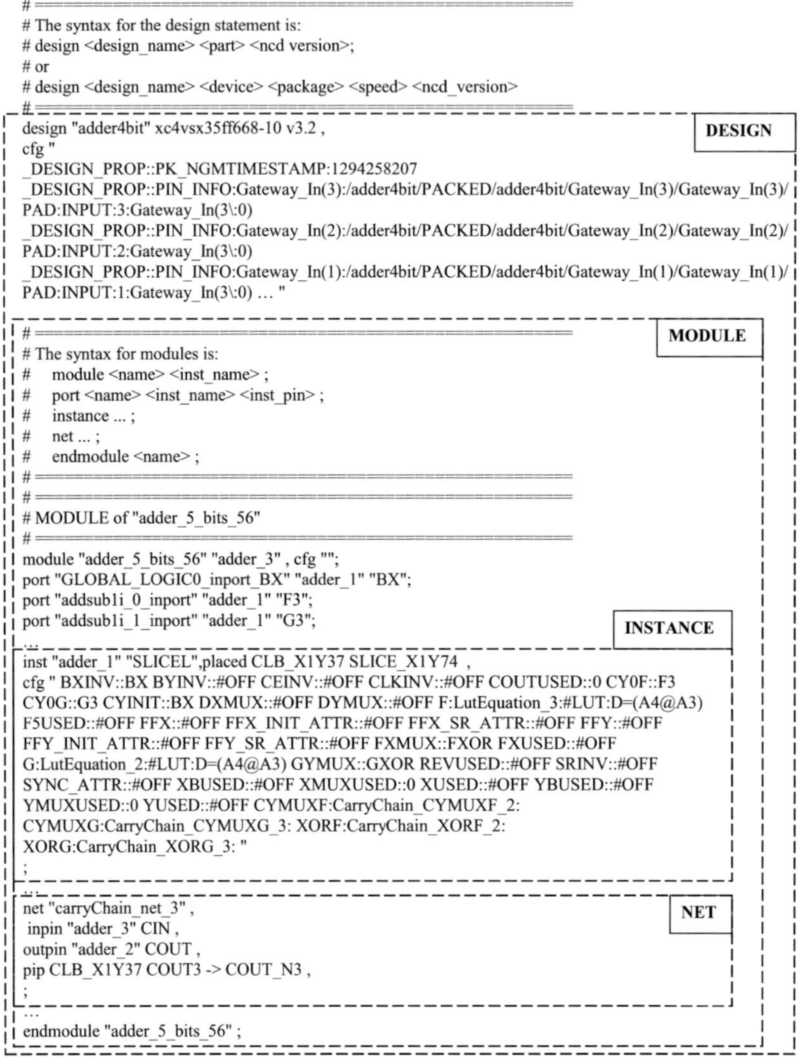

```
# ================================================================
# The syntax for the design statement is:
# design <design_name> <part> <ncd version>;
# or
# design <design_name> <device> <package> <speed> <ncd_version>
# ================================================================
design "adder4bit" xc4vsx35ff668-10 v3.2 ,                          DESIGN
cfg "
_DESIGN_PROP::PK_NGMTIMESTAMP:1294258207
_DESIGN_PROP::PIN_INFO:Gateway_In(3):/adder4bit/PACKED/adder4bit/Gateway_In(3)/Gateway_In(3)/
PAD:INPUT:3:Gateway_In(3\:0)
_DESIGN_PROP::PIN_INFO:Gateway_In(2):/adder4bit/PACKED/adder4bit/Gateway_In(2)/Gateway_In(2)/
PAD:INPUT:2:Gateway_In(3\:0)
_DESIGN_PROP::PIN_INFO:Gateway_In(1):/adder4bit/PACKED/adder4bit/Gateway_In(1)/Gateway_In(1)/
PAD:INPUT:1:Gateway_In(3\:0) ... "

# ================================================================            MODULE
# The syntax for modules is:
#     module <name> <inst_name> ;
#     port <name> <inst_name> <inst_pin> ;
#     instance ... ;
#     net ... ;
#     endmodule <name> ;
# ================================================================
# ================================================================
# MODULE of "adder_5_bits_56"
# ================================================================
module "adder_5_bits_56" "adder_3" , cfg "";
port "GLOBAL_LOGIC0_inport_BX" "adder_1" "BX";
port "addsub1i_0_inport" "adder_1" "F3";
port "addsub1i_1_inport" "adder_1" "G3";
                                                                    INSTANCE
inst "adder_1" "SLICEL",placed CLB_X1Y37 SLICE_X1Y74 ,
cfg " BXINV::BX BYINV::#OFF CEINV::#OFF CLKINV::#OFF COUTUSED::0 CY0F::F3
CY0G::G3 CYINIT::BX DXMUX::#OFF DYMUX::#OFF F:LutEquation_3:#LUT:D=(A4@A3)
F5USED::#OFF FFX::#OFF FFX_INIT_ATTR::#OFF FFX_SR_ATTR::#OFF FFY::#OFF
FFY_INIT_ATTR::#OFF FFY_SR_ATTR::#OFF FXMUX::FXOR FXUSED::#OFF
G:LutEquation_2:#LUT:D=(A4@A3) GYMUX::GXOR REVUSED::#OFF SRINV::#OFF
SYNC_ATTR::#OFF XBUSED::#OFF XMUXUSED::0 XUSED::#OFF YBUSED::#OFF
YMUXUSED::0 YUSED::#OFF CYMUXF:CarryChain_CYMUXF_2:
CYMUXG:CarryChain_CYMUXG_3: XORF:CarryChain_XORF_2:
XORG:CarryChain_XORG_3: "
;

net "carryChain_net_3" ,                                            NET
 inpin "adder_3" CIN ,
outpin "adder_2" COUT ,
pip CLB_X1Y37 COUT3 -> COUT_N3 ,
;
...
endmodule "adder_5_bits_56" ;
```

Figure 6.13: Abbreviated version of an XDL file [50]

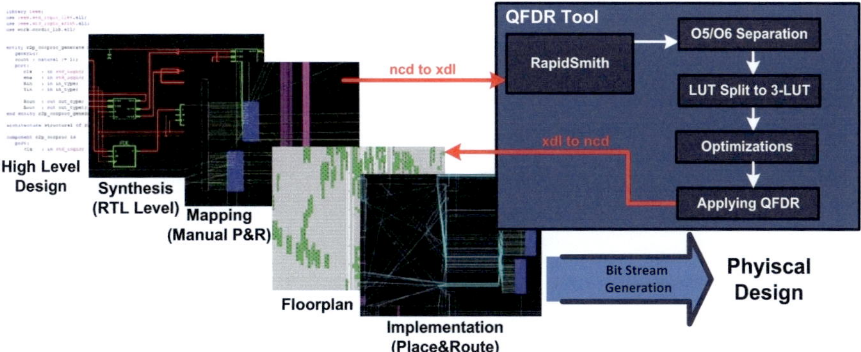

Figure 6.14: Integration of QFDR in ISE CAD tool using Rapidsmith. The ncd format of mapped design is converted to xdl and is presented in an objective form in RapidSmith. QFDR tool uses this objective form and modify the mapped design. Finally it creates a modified ncd file.

6.7.1.3 Net Statement

Net statements present the connections between LUTs in every slice instances. In order to complete QFDR, connections between Forcer and Decider LUTs must be modified. It includes adding new nets for new LUTs and also modifying the connections between existed LUTs. Every net has an outpin and several inpins.

6.7.2 RapidSmith

In order to automatically add QFDR to the design, the RapidSmith framework [103] was used. RapidSmith is a JAVA based API built to manipulate XDL files and is used to integrate QFDR in Xilinx ISE.

RapidSmith was extended with the proposed QFDR approach. Due to the flexibility of this framework, every extension on xdl form can be done to achieve fault tolerance targets.

In order to apply QFDR, instance statements in xdl format are the key statements. Because through them, it is possible to access the function of every LUT and modify it. Figure 6.14 shows the process of integrating QFDR in synthesis flow using RapidSmith. In the first step the design is transformed and parsed to the internal object representation. In order to apply this methodology some preprocessing of the design must be made.

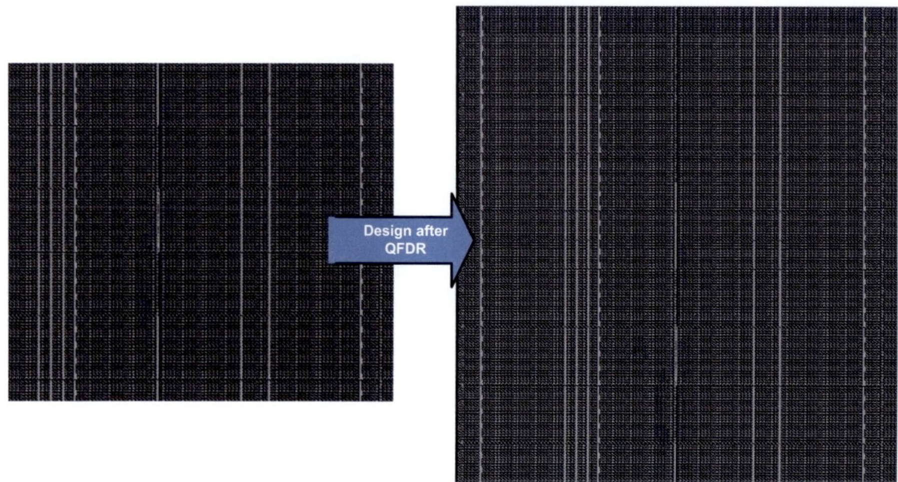

Figure 6.15: A Cordic implementation before and after applying QFDR on FPGA Editor. Placement of Slices has done without using any optimized algorithm.

In the first step, the instance statements are read and different kind of slices (SliceM and SliceL) are separated. In every kind of slices, the whole process of QFDR modification is done. Firstly, LUTs which use both O5 and O6 are divided to two LUTs. Then, using Shannon decomposition method, LUTs with more than 3 inputs are detected and reduced to 3 inputs. Afterwards, optimizations are done. All this is done in using the object structure of RapidSmith. Then the logic is quadruplicated in the next step and the functions are modified to realize the force decide behavior. Thereafter the wiring is updated to connect the different levels of the quadruple force decide structure. Finally the placement is optimized.

Some modifications also are needed in net statements in order to implement the wiring in QFDR. In net statement, inpins must be also duplicated and reconnected to the corresponding Quadded LUT.

The whole process is done on the XDL format using RapidSmith which presents the structure in objective form. Using RapidSmith, Slices and Nets can be used as objects and modification can be done on objects. Finally the modified objects can be represented again in XDL.

6.8 Integration of QFDR in Future FPGA architectures

Although QFDR can easily be implemented in current FPGAs, future implementations can still be improved in various ways to decrease area overhead and increase reliability.

As described, QFDR applies two basic rules to the duplicated inputs of every LUT: Force and Decide. Currently, these rules are applied to a design by changing the function which is implemented on LUT. To cope with the extra area overhead that was mentioned in the previous section, one solution is to adjust every Slice structure to QFDR in future FPGA designs.

As shown in Figure 6.11, the structure of a Forcer and Decider is easily implementable in hardware. Both include a comparison between duplicated inputs in addition with an AND or OR gate depending on the Force or Decide roles.

In this way, every LUT will be modified to two different sorts of LUTs: Forcer LUT or Decider LUT. Both LUTs will basically have the structure of current LUTs. In addition some new hardwares in their input lines needs to be added.

In Forcer LUT, all duplicated input lines are XOred in order to recognize any difference in the duplicated inputs. If it occurs, the output is forced to '1'. This is done using the OR at the output of XORs. In a Decider one, an AND gate will be added to the input lines of LUT to decide for the correct input value.(Figure 6.16).

The only information which is needed is the role of a function which might be Forcer or Decider, and then, the original design can be easily fitted on the suitable LUT based on standard place and route algorithms.

In addition this would extremely simplify the QFDR integration process. Firstly, decomposition of functions is not the case any more, because six (or more) inputs can be used in design. The reason is that, duplication is done in a prior level to the inputs of described AND and OR gates.

It will also extremely decrease the area overhead. If one assumes a three input LUT which is QFDRed in the current manner, the number of transistors for a K-LUT (N_K) which are used can be computed as follow:

$$N_K = 2^K \times 6 + N_{K-MUX} \qquad (6.2)$$

Here, 2^K is the number of SRAM cells which are needed to realize a K-LUT function, 6 is the number of transistors in a SRAM-cell [114][77], and N_{K-MUX} is

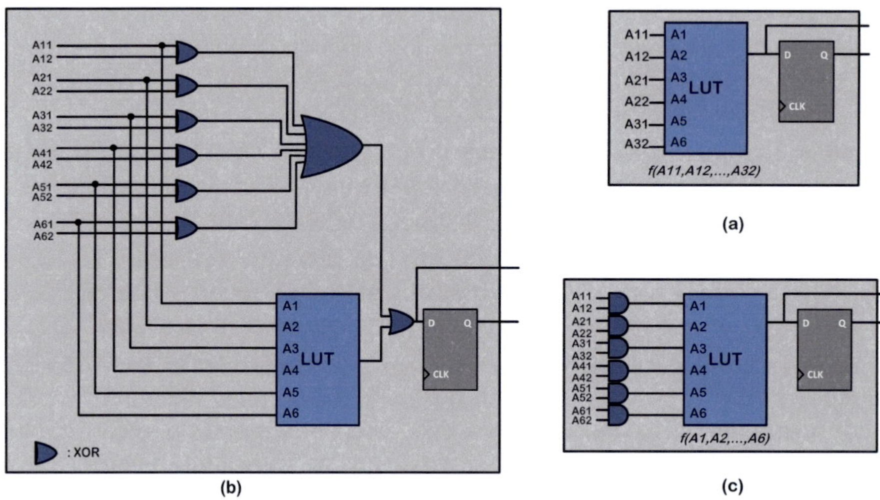

Figure 6.16: (a) Current LUTs which are modified for QFDR. Here three inputs can be used in the function and these inputs are duplicated. (b) A new implementation of a LUT in adidition with extra hardware which realizes a Forcer. Duplicated inputs force the output to one in case of difference in duplicated inputs. This is done using an external hardware.(c) A new implementation of a LUT in adidition with extra hardware which realizes a Decider. Inputs are externally duplicated and ANDed together and the result is connected to LUT.

the number of transistors which are needed to realize the multiplexer in K-LUT. In case of three input LUT, although K is equal to three, but due to QFDR K is assumed 6 for a 3-LUT.

If the new hardware structure is used, K can be assumed 3. Number of transistors in extra hardware for the Forcer LUT will be:

$$N_{K_{Forcer}} = N_K + K \times N_{XOR} + N_{OR} \tag{6.3}$$

In case 3-LUT, K is equal to 3 and N_{XOR} and N_{OR} are respectively the number of transistors in XOR and OR gates.

In conclusion the number of transistors in extra hardware for the Decider LUT will be:

$$N_{K_{Decider}} = N_K + K \times N_{AND} \tag{6.4}$$

Table 6.2 shows an estimation of the difference for different LUTs.

Table 6.2: Different LUTs number of transistors for current QFDR method and the hardware integrated one.

	QFDRed LUT	Forcer LUT	Decider LUT
1-LUT	30	20	18
2-LUT	114	48	42
3-LUT	414	84	78
4-LUT	-	144	138
5-LUT	-	252	246
6-LUT	-	456	450

Secondly, function modification is not more necessary. Currently function modification is done, because the behavior of AND and OR gates must be realized in the function. If these gates exist separately, function does not need to be modified any more.

Integration of QFDR in hardware will also reduce the SRAM cells which are needed to configure the QFDRed design on FPGA. Currently, modified functions are needed to be configured on LUTs. However, if the behavior of QFDR exists in hardware, the number of SRAM cells which are needed to configure the design, will be reduced to the original design in addition to the routing modifications.

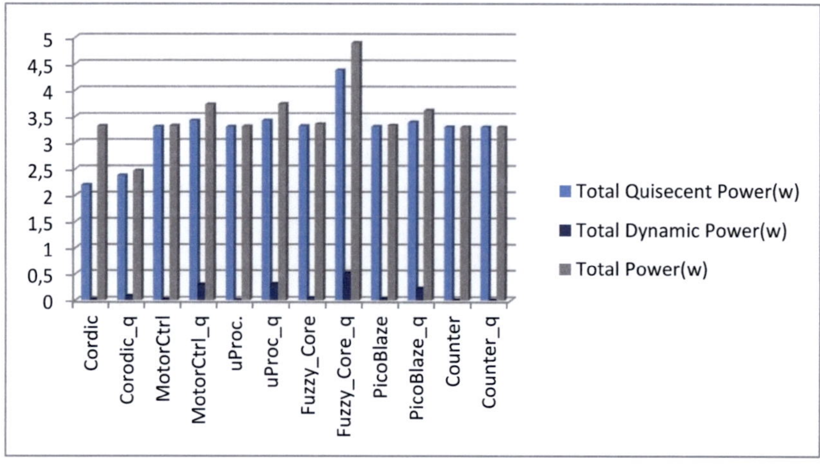

Figure 6.17: Static and Dynamic Power usage of different benchmarks and their QFDR form.

6.9 Power Estimation

In this section, several circuits are compared based on the selected XC5VLX110T Virtex-5 FPGA in term of power usage.

There are two main groups of power consumption during operation mode in SRAM-Based FPGAs: Dynamic and Static. The peak in power consumption in SRAM-Based FPGAs is caused by the fact that the logic states are not determined [95].

However, this increases when the overhead of QFDR is inserted to the architecture. While the static power usage in SRAM-based FPGAs can be ignored [95], dynamic power usage shows a few overhead in case of QFDR. Power analyzing is done using XPower Analyzer which is provided by Xilinx ISE [9]. In Figure 6.17 for several benchmark the comparison between a benchmark power usage and its quadded form power usage is done.

6.10 Fault Tolerance

Using the simulation tool, faults are inserted in different signals randomly during their simulations. Every QFDRed benchmark is executed 1000 times per every executions 5, 10, 20, and 50 faults are periodically inserted. The results of Fault tolerance are shown yn Figure 6.20.

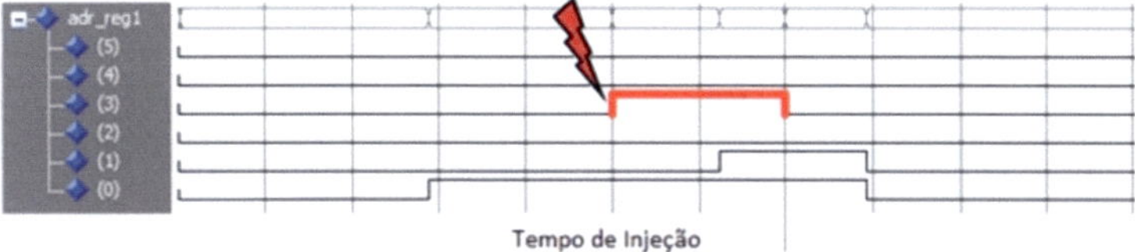

Figure 6.18: An SET which is simulated on Signal by fault injection

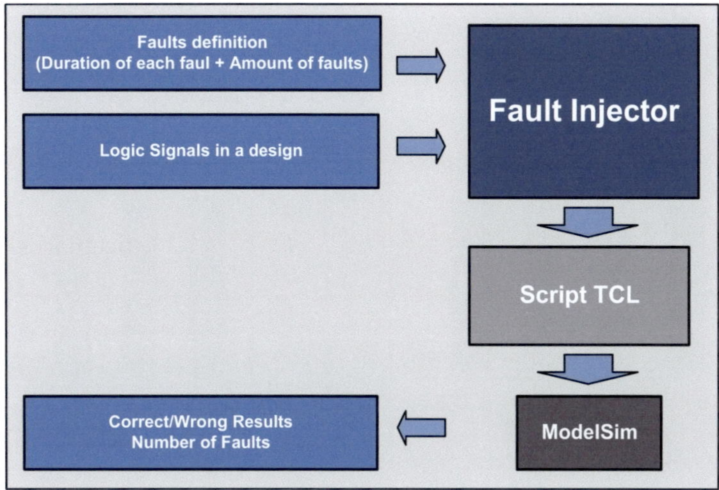

Figure 6.19: The structure of Fault Inejctor

In order to measure the fault tolerance of every structure, every circuit is simulated several times by changing the value of the LUTs which is a simulation of SET affections. The simulation of radiation effect is done using the ModelSim simulation tool.

Design developments with fault tolerance targets require a test phase which is composed fault injection. Ideally, the system is irradiated with energized particles. However, this is a process that requires appropriate facilities, use of a custom test adapter and control equipment, and extracting data of the project under test, which this results in high cost. To circumvent this problem, a simulation framework is used which has been integrated into ModelSim [7] and is capable of generating and injecting faults in a design.

ModelSim provides both a graphical environment for simulation as a console for writing scripts in the language Tool Command Language(TCL) and the output in text form. Beside these facilities, a feature which is interesting for the simulation

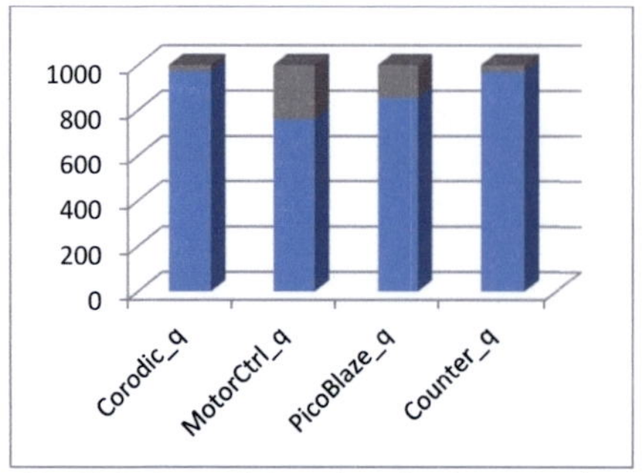

(a) 50 faults are injected randomly per each execution.

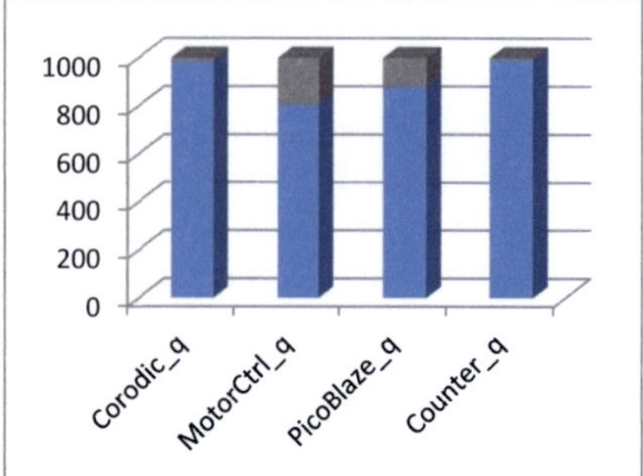

(b) 20 faults are injected randomly per each execution.

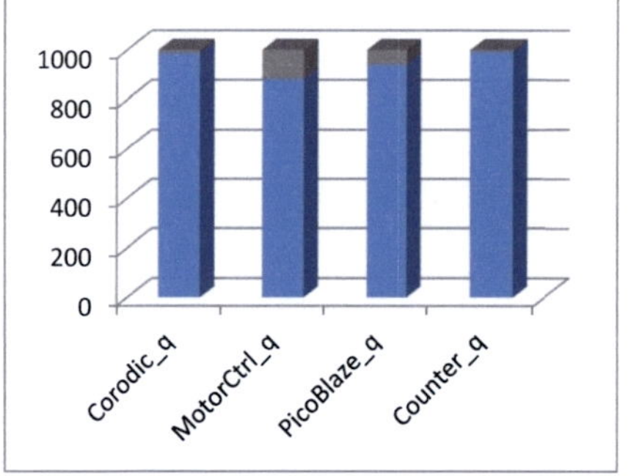

(c) 10 faults are injected randomly per each execution.

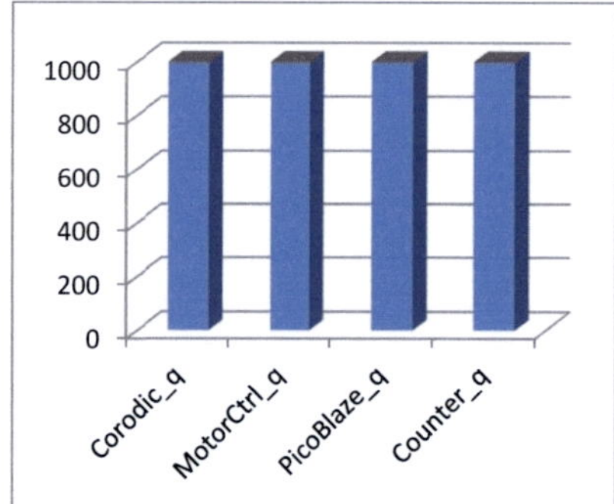

(d) 5 faults are injected randomly per each execution.

Figure 6.20: 1000 times executions. Different number of faults which are injected in every execution time.

tool is the ability to access, read, and write any logic signal exists in the simulation process at any desired time. The simulation framework, which is used here, uses this feature and injects faults on logic signals of the designs.

The fault injector's main object is to generate scripts in TCL format to run on ModelSim. While the fault injection is on LUTs, the fault injector provides a good estimation of fault coverage.

The fault injector is divided into two distinct stages. 1) Generating a set of faults. 2) Simulating the fault set using ModelSim and collecting the results generated by ModelSim. Figure 6.19 shows the process.

The faults can act as an SET or SEU. They can be transient or permanent. This is defined in the TCL script for each fault. An SET can behave like in Figure 6.18.

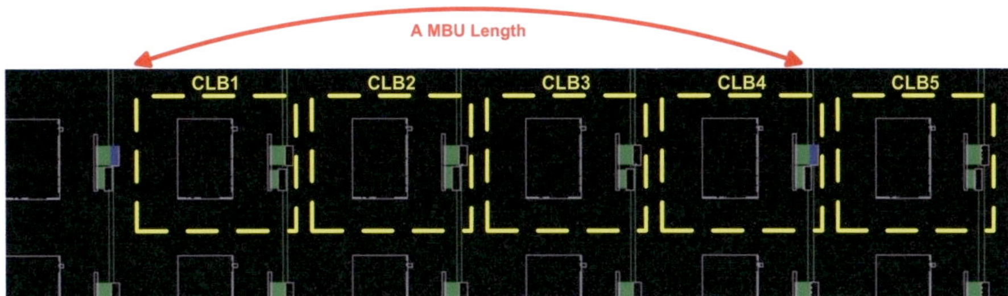

Figure 6.21: One 3 bit MBU can effect a length of 4 CLBs based on experiments on [101]. Mapping of the Slices of the same fine grains should be done in a distance of 4 CLB to protect the design against MBUs.

6.11 MBU Mitigation

MBUs are increased in new FPGA technologies since the technology scales are decreased. Multiple Bit Upset can be a result of a strong strike or a single upset which flips its neighbors. In both cases it is possible, that more than one copy in a quad is affected and due to this fact, some modification is needed in order to mitigate against MBUs.

In order to make the design on Virtex-5 FPGAs tolerant against MBU, as described in the previous chapter, one method is to change the placement of the quadruplicated LUTs.

As described in Chapter 3, the measurement shows that currently the four bit upsets are the most crucial MBUs which can be detected (Table 3.3). If an MBU of four bits occurs in an FPGA, it is possible that a distance of 4×4 CLBs, is affected. However, this distance can flexibly change because the probability of 1 SEU or 2 MBUs are much more as results show.

In Figure 6.21 a possible placement of slices on FPGA is shown. The key issue of the reliable placement is that, slices are placed in a distance which an MBU can cover.

This means, if a design is QFDRed, every slice has three other copies. If the four copies are neighbor slices, it is possible that after an MBU, all the four slices are affected and none of the copies functions any more. One idea is to place 4 copies in a distance of 4×4, depending on the MBU rates. This is applicable to all fine grain fault tolerance methods which are applied on FPGA homogeneous slices. However, long distance placement will affect the routing network. More nets are needed to realize the routing. This affect power usage and performance of the system.

Table 6.3: Number of PIPs in different designs does not before and after placement modifications

Design	Number of PIPs (before and after placement modification)
Cordic	2211
Counter	74
uProc.	5837
FuzzyCore	14281

In order to measure the overhead of placement modifications, the structure of routing was analyzed. Router uses algorithms to find optimal way between two placed slices. Programmable Interconnection Points or PIPs are programmable boxes which are between slices and are used to connect two or more slices together. If two slices are placed far from each other, normally more nets are need to connect slices though PIPs together. This results in a more complicated routing in the design which affect dynamic power usage.

However the number of PIPs does not change ([27]) . Although a quad four components are placed far from each other, the other components form other quads may be placed in between and the final number of PIPs remains similar. Table 6.3 shows the number of PIPs for some benchmarks. However, the number of nets which connect two slices together are increased. The reason is that more PIPs are needed to connect two slices as they are placed far from each other.

Recent studies have characterized different bit errors arising from an SEU and suggests that 1 to 5 % of the SEUs can cause multiple bit upsets.

Based on [101], the probability of adjacent double bit errors is much higher than other multiple bit errors. However, modification of reliability tools to tolerate against MBUs is necessary in near future.

Coping with the MBU problems in more details includes the future works and is not in the scope of this contribution.

7 Exploring Quadded Logics for Reliability in Future Technologies

In the previous chapters, the advantages of reconfigurable architectures such as FPGAs in space and their obstacles were described. The focus was on SRAM based FPGAs which use current CMOS technology and its benefits.

Gordon Moore predicted in 1965 that the number of transistors which could economically be placed on an integrated circuit would increase exponentially with time. Eventually this leads to some physical limitations in CMOS technology and the need for new architectures are risen up. In this chapter, the focus is turned around the nano architectures which are expected to be the future architecture which is one of the best alternative to CMOS. The programmability issue in state of the art nano architectures is mentioned and its challenges are discussed.

7.1 Nano Architectures and their Challenges

In these years, nano architectures have emerged as an alternative to CMOS technology for the integration within CMOS as one approaches the end of the semiconductor road map and lithography based fabrications meet physical limitations [102][115][75]. Now it is becoming possible to look beyond lithography and explore how devices can be built without relying on lithography to pattern the smallest feature sizes [39][37] [40].

Bottom-up self-assembly techniques are demonstrated to define key feature sizes [39][118][62]. Self-assembly is a process in which molecules adopt a defined arrangement without guidance or management from an outside source. This noramlly results in regular structures. It has been shown that, by using bottom-up self-assembly techniques, it is possible to build nano devices such as carbon nanotubes and silicon nanowires. These devices have been rated among the most promising of all the new technologies under investigation [12].

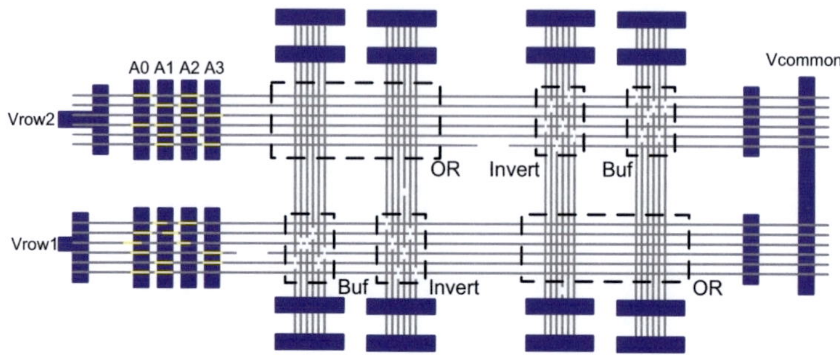

Figure 7.1: Basic structure of nanoPLA with two planes. Broken wires can be distributed overall the PLA. It consists of two ORs (Top-Left and Bottom-Right) and Buffer and Inverter on the other side of every plane [42].

These techniques allow to build features which are just a few atom scales. It is projected that, only regular structures such as the 2D crossbar can be manufactured. Chemically self-assembled structures, as the building blocks for molecular scale computing, are by their nature very regular. This way regular programmable architectures are currently being investigated in research. These architectures share some similarities with conventional FPGAs. In particular they are well suited to be used to implement regular arrays similar to FPGAs [29][37].

In general, by reducing the feature size, the control over the fabrication process is reduced as well. There is no outside control on self assembly processes and this makes it unlikely to construct complex circuits without any defect [29]. In any kind of nano architectures, nanowires are a few atoms long in the diameter. The contact area between nanowires contains only a few tens of atoms.

Even in the largely manufactured conventional lithographic systems, the technology is facing reliability challenges. As a result, design rules will change and become more restricted to the regular design structures [113].

Furthermore, circuit designers can no longer design simply by technology design rules and expect a functional and still scalable design. Designers must know when to use more relaxed rules and not simply relax the rules on the entire design, which negates physical scaling[20].

Hence, nano scale architectures come with a new set of challenges due to defects as a result of smaller cross section and contact areas. Designs in this scale must be defect tolerant, making the test or characterization for usability an essential step in building circuits with nanowires [87][42].

For emerging nano technologies the defect rate is even higher due to the small feature size and bottom up nature of the design. Although the defect rate for

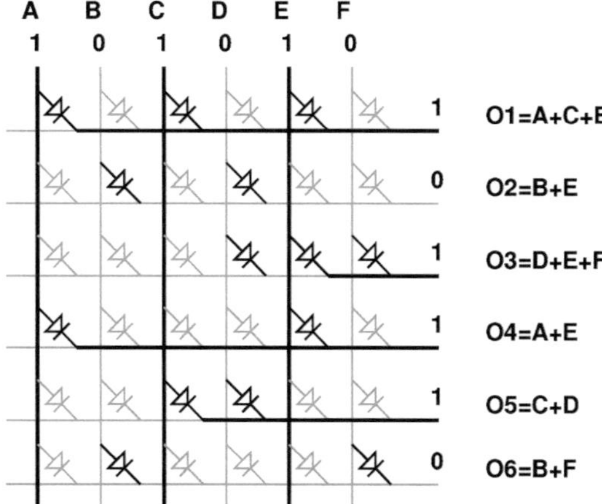

Figure 7.2: Wired-OR plane operation. Programmed on crosspoints are shown in black; off crosspoints are shown in grey. Dark lines represent a NW pulled high, while light lines remain low. Output NWs are marked dark, starting at the diode that pulls them high, in order to illustrate current flow; the entire output NW would be pulled high in actual operation [40].

emerging technologies is expected to decrease once the technologies are more mature, due to the nature of the fabrication process it is still expected to be high [35][59].

7.2 NanoPLA Architecture

Several researchers have begun to explore programmable logic structures in this scale. [60] introduces a vision for this kind of molecular scale logic. Nano Fabrics [37] are an example of this vision. They use two terminal diode crosspoint nanowires. [40] has explored how to use nanowires to build sub-lithographic PLAs and interconnected PLAs which is called nanoPLA. They built a two plane PLA with decorated silicon nanowires and device building blocks.

nanoPLAs are the most promising circuit solutions because of their regular geometry which enables large scale fabrications in order of nanometers. The regular nature of these devices can support implementation of Programmable Logic Arrays (PLAs). This means that, in future they seem to be the best alternative of reconfigurable architectures like FPGAs.

Figure 7.3: Crossbar routing configuration. Programmed on crosspoints are shown in black; off crosspoints are shown in grey. The crossbar is programmed to connect $A \longrightarrow T$, $C \longrightarrow V$, $D \longrightarrow S$, $E \longrightarrow U$, and $F \longrightarrow R$ [40].

As Dehon firstly suggested in [42], NanoPLAs, like conventional PLAs, consist of two programmable NOR planes (Figure 7.1). Each one of them consists of two arrays: a logic array and a buffer/inverter array.

The logic array is the programmable part of the NOR plane. Using junctions which are bistable crosspoints, the logic array implements the OR function of its inputs [87]. If there is a way to program the crosspoints into high or low resistance states, it is possible to program OR logic into a crosspoint array.

One can consider a single row of nanowire. If any of the columns of nanowires which cross this row are connected with low resistance crosspoint junctions and are driven to a high voltage level, the current into the nanowire in column will be able to flow into the row and charge the row up to a higher voltage. If none of the connected columns is high, the row nanowire will remain low.

Consequently the row nanowire effectively computes the OR of its programmed inputs. When an input participates in an OR function of the design, the junction of the input and the wired-OR will be closed while applying the high voltages to the nanowire which crosses the junction and if it does not participate, will be left open as it is in the initial state. Figure 7.2 shows the wired OR plane operation.

If one restricts himself to connecting a single row wire to each column wire, the crosspoint array can serve as a crossbar switch. This allows to route any input (column) to any output (row). Figure 7.3 mentions an example for crossbar routing configuration.

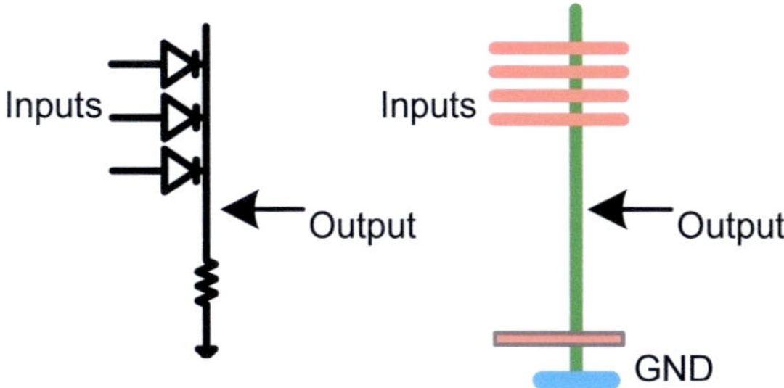

Figure 7.4: (a) Wired OR logic in crosspoints. (b) Its corresponding structure in nanoPLA

Each of the connected junctions behaves like a diode and produces the wired OR logic of its inputs. If any of the inputs is high, it pulls up the OR term output. Figure 7.4 shows wired OR circuits, which can be constructed by programming the crosspoints.

As Figure 7.1 shows, the NOR plane also consists of a buffer/inverter array. Buffer/inverter arrays, which have nonprogrammable junctions [42][41], are used to restore/invert the input signals. In order to provide a NOR function, the OR-term signals use these arrays for inversion. The output of the inversion can then be buffered for further usages.

The inputs of the buffer/inverter array are the OR-term nanowires or the primary input nanowires. The outputs are always the restored signals. The restored signals either are buffered or inverted. Accordingly, the input signals to the logic array can be buffered or inverted. Some of the signals are just buffered while some are only inverted, and some are both buffered and inverted [87].

For example, to implement the XOR function on nanoPLA, every function must be rewritten in NOR form to be able to be programmed on nanoPLAs (See Figure 7.6 in section 7.5).

7.3 Crosspoint Arrays and Defects

As already mentioned, due to atomic scale of nanowires, defects are very common in nanoPLAs. Figure 7.1 shows possible defect places in nanoPLA which are used in the defect model. Some spaces in nanowires are shown to be broken. Two main defects in nanoPLAs are defects in programmable crosspoints due to the structure of the junction and defects in nanowires [42].

[87] has distributed a probability model for the crosspoint defect. Authors show, Nanowires with 40nm diameters have the cross sectional area of $1600nm^2$ and as a result 1100 programmable molecules can be placed in the cross section. If the nanowires of width 8nm which is demonstrated in the later technologies is used, the cross sectional area will be $64nm^2$ which is $\frac{1}{25}$ of the first cross sectional area and scaling the number of molecules in the cross section with $\frac{1}{25}$ yields about 44 molecules. When P is the probability of a single molecule on a unit of area, the probability distributed function of the number of molecules on a cross sectional area of 44 times molecule area is:

$$P = \binom{44}{x} P^x (1 - P)^{44-x} \tag{7.1}$$

x is the number of molecules in the cross sectional area. If N_{min} is the minimum number of molecules in the cross section to make a junction programmable, the probability of a junction to be programmable is about 0.85. [86] has illustrated the above distribution function in addition with its Cumulative Distribution Function with P=0.8. Figure 7.5 shows these two curves.

[59] reported that 95% of the wires measured had good contacts, and [35] reported that 85% of crosspoint junctions measured were also properly usable. [105] mentioned that both of these experiments were early measurements and therefore, the yield is expected to be improved. [87] reported about more than 90% good nanowires and more than 80% defect free crosspoint junctions. In the approach the system will work properly even if only 80% of nanowires and junctions are good.

But if the defect rates are reduced in the future some new methodologies such as the proposed one are needed for reliability. In nano architectures in general, variation effects on design result in a higher rate of soft errors, due to the intensive structure of carbon nano tubes. Therefore, it is mandatory to implement high-reliable designs [21].

7.4 Conventional Test and Characterization process in nanoPLAs

[40] presents an exhaustive test procedure for nanoPLAs. The test procedure is done before programming a function (for example XOR) on nanoPLA. Test procedure includes two steps: *Discovering Present Addresses* and *Discovering Polarities*. After knowing about the presence and polarity of each address, the *Program-*

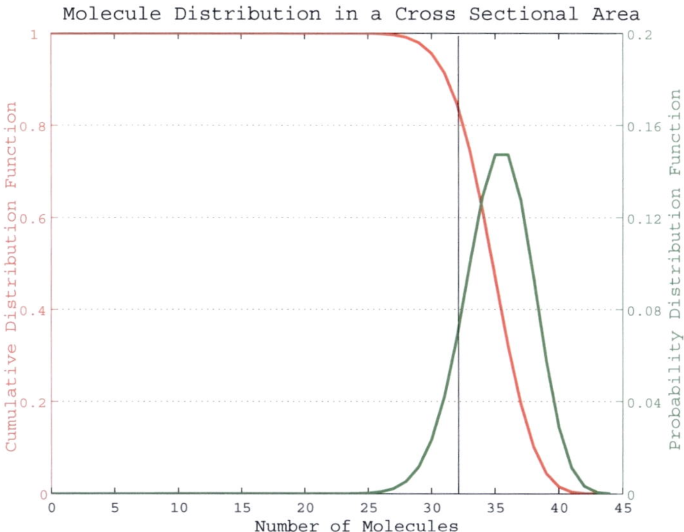

Figure 7.5: Cumulative Distribution function and the probability of distribution function [86].

ming Diode Crosspoints procedure is performed to configure NOR-NOR planes to implement the corresponding logic function. More details on programming crosspoint junctions can be found in [40].

In this methodology, defects in nanowires can be detected by applying a voltage to only one nanowire through the decoder, then reading the value from the other end and determining if the nanowire conducts across its length. The horizontal nanowires can be tested by applying a voltage through the decoder from one end and read back the value from the other end.

To test the vertical nanowires the high voltage is applied to a horizontal nanowire through the decoder, then the signal controls a vertical nanowire current through the FET-like restoring junction of the buffer/inverter array. The value of the vertical nanowire can be read from the other end.

As [105] also addressed, the test and characterization procedures take together about 128 steps for each nanoPLA. This leads to high test cost when considering large scale production.

In the work, this process is removed and instead of it, the function structure is modified to make it tolerant against defects on nanowires. It means that instead of doing a process to find suitable (not defected) nanowires, which consumes time, a defect tolerance function on existing nanowires is programmed.

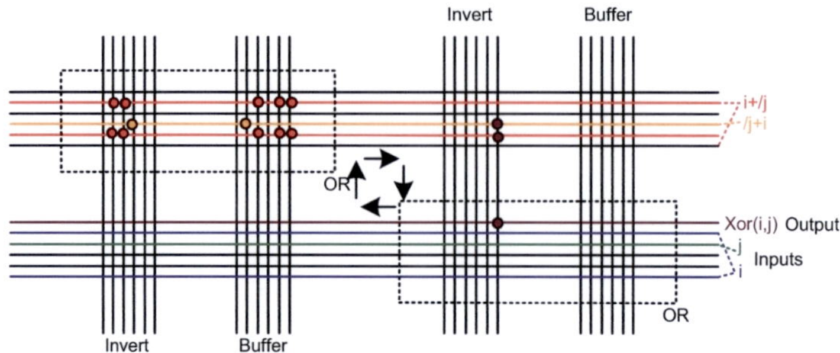

Figure 7.6: XOR implemented on NanoPLA

In order to program a function on nanoPLAs, it must be reconstructed in NOR gates. This property is suitable for quadded NOR logics, which use the behavior of NOR gates and wires them together making the function tolerant against faults.

7.5 Quadded Logic on nanoPLA

As described in section 7.2 the complete characterization for each nanoPLA is done in the conventional test to determine the exact programming steps that need to be followed to program the intended function. Although this approach can provide exact function and prevent extra spares and due to this guarantee a good yield, the huge number of characterization steps may not be suitable for high demand fabrication process.

[105] presents a different paradigm of testing which is based on removing the process of characterization and inserting spare wires to guarantee the correct function in case of defect in original wires.

In the previous chapters the fundamentals of quadded logics and their usages were described. A methodology based on quadded logics is introduced, which makes not only a defect tolerant design but also reliable even against other kind of transient errors. The basic idea is to use quadded NOR logics in nanoPLAs and make the design defect and fault tolerant. NanoPLA is especially suitable for quadded NOR form because NORs can be provided using nanoPLA.

First, the fundamentals of quadded NOR logics in nanoPLAs using the wrapping feature of nanoPLAs is described. This feature makes it easy to implement function on nanoPLAs.

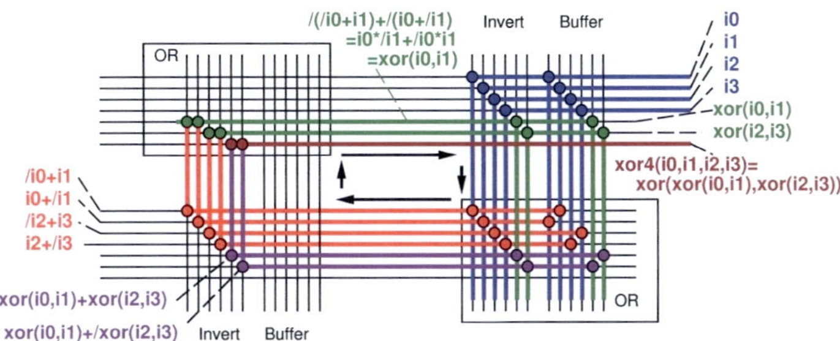

Figure 7.7: Wrapping feature in nanoPLAs is used here to implement four inputs XOR on nanoPLA. Two input XORs are generated in the first wrap and four input one in the second wrap [42].

To introduce the concept of quadded logics for nanoPLAs, XOR logic which is rewritten in NOR logics is shown in Figure 7.8 in both its non-redundant and quadded forms. XOR is presented in NOR form due to the nanoPLA usage purpose.

Each NOR gate in Figure 7.8(a) is replaced by four NOR gates in Figure 7.8(b). Each quadded form NOR has twice as many inputs as the non-redundant gate. Four outputs of each stage are divided into two sets of two outputs. Each set provides inputs to two gates in the succeeding stage.

This arrangement of NOR gates and interconnections provides error correction capability. Error definition varies from bit flipping to defective wires which result in bit flipping. Both are probable in nanoPLAs. Any single error and a number of multiple errors introduced in quadded XOR will not cause failure in the functioning of the system.

To illustrate this capability, in Figure 7.8(b) assume that i_0 generates '1' instead of '0' as the result of defect in wire. The result of neighbored NORs (S_2 and S_3) is one as should be and i_0 will effect S_0 and S_1 to '0'. Because '1' is the dominant value for NOR logic, it will remove this error in the next stage using the correct value of the neighbors. More details on quadded logics and their proof can be found in Chapter 5 and [90][70][120][121].

7.5.1 Multilevel logic evaluation in nanoPLA

As described in the previous section, every function must be rewritten to NOR parts to be able to map on nanoPLA. The XOR in NOR form is rewritten and is programmed on nanoPLA. In order to use quadded NOR logics, one must have the entire design in NOR gates. It is implemented as shown in Figure 7.8(a).

Equation 7.2 shows the NOR formed XOR logic which is used to map XOR on nanoPLA in Figure 7.6. Implementation is done after passing test and characterization process. For simplicity, in the continuing figures, the nanoPLA is shown by its programming part in addition to buffer and inversion parts.

$$O = i \oplus j$$
$$= i.\bar{j} + \bar{i}.j$$
$$= \overline{\overline{(\bar{i}+j)} + \overline{(i+\bar{j})}} \tag{7.2}$$

i and j are inverted and buffered in the below plane right and are programmed as corresponding OR which is needed in the up plane. Then they are inverted in the top plane and finally are ORed again in the bottom plane as the result.

Rather than using a separate physical plane for every logic stage of evaluation in a spread PLA mapping, the logic function can be wrapped around the PLA multiple times [42].

Consider a 4-input XOR. The XOR is rewritten to some NOR terms. [42] shows that it is possible to wrap the XOR twice through the PLA, computing 4 input XOR as a cascade of two level XORs. In Figure 7.7, implementation of the 4-inputs XOR on nanoPLA using wrapping feature has been shown [42].

Figure 7.6 illustrates the original implementation of NOR-NOR form of XOR. The inverting and buffering parts of nanoPLA are shown. In the bottom row, i and j as inputs are derived and inverted or buffered as shown. In the top lines corresponding ORs are generated and finally in the top left line the final inversion is done and the XOR result is generated.

The multilevel programming ability, which is presented in Figure 7.7, for a 4-input XOR logic wraps around the nanoPLA in order to program the sub XORs and the main XOR and the end as follows:

$$O = i_0 \oplus i_1 \oplus i_2 \oplus i_3$$
$$= (i_0 \oplus i_1) \oplus (i_2 \oplus i_3)$$
$$= \overline{\overline{((i_0 + \overline{i_1}) + \overline{(\bar{i_0} + i_1)})} \oplus \overline{((i_2 + \overline{i_3}) + \overline{(\bar{i_2} + i_3)})}} \tag{7.3}$$

Every part is mapped on nanoPLA and used in the next wrapping. This feature is used in the next section while the quadded XOR is mapped on nanoPLA.

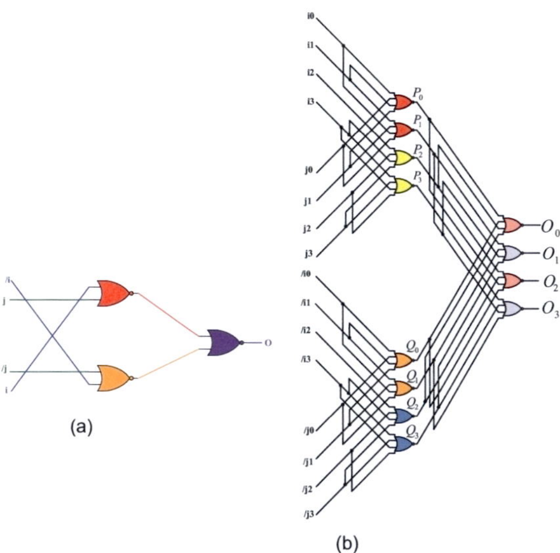

Figure 7.8: (a) XOR logic in NOR form. (b) Quadded NOR Logic form of XOR

7.6 The Proposed Approach

The multi-level logic map capability in nanoPLA is used to map quadded logic form on it and make it tolerance against defects.

The nanoPLA, illustrated in Figure 7.1, is the target architecture for design methodology. In Figure 7.6 is shown how an XOR is programmable on nanoPLA, after passing the process of test and characterization of nanoPLA.

The programming strategy, implements the quadded form XOR instead of XOR on nanoPLA. Every nanoPLA has two planes and every plane consists of 6 wires. The number of nanowires in every plan is increased in order to achieve the quadded form. Based on [52] there is no physical limitation to increase the number of nanowires in every plane. Because of the nature of quadded logics, which quadruplicates the input numbers, the number of nanowires in nanoPLA is increased to be further used in quadruplicated form.

Equation 7.2 rewrites the XOR in NOR forms to be suitable for nanoPLA. The quadded form XOR in NOR gates is modified based on the Figure 7.8. There is still NOR formed, but the number of inputs is quadruplicated as a result of quadded modification. Equation 3.1, 7.5 and 7.6 show the rewritten quadded XOR in quadded form.

$$P_0 = \overline{(i_0 + i_1 + j_0 + j_1)}$$
$$P_1 = \overline{(i_1 + i_0 + j_1 + j_0)}$$
$$P_2 = \overline{(i_2 + i_3 + j_2 + j_3)}$$
$$P_3 = \overline{(i_3 + i_2 + j_3 + j_2)}$$

$$(7.4)$$

$$Q_0 = \overline{(\overline{i_0} + \overline{i_1} + \overline{j_0} + \overline{j_1})}$$
$$Q_1 = \overline{(\overline{i_1} + \overline{i_0} + \overline{j_1} + \overline{j_0})}$$
$$Q_2 = \overline{(\overline{i_2} + \overline{i_3} + \overline{j_2} + \overline{j_3})}$$
$$Q_3 = \overline{(\overline{i_3} + \overline{i_2} + \overline{j_3} + \overline{j_2})}$$

$$(7.5)$$

$$O_0 = \overline{(P_0 + P_3 + Q_0 + Q_3)}$$
$$O_1 = \overline{(P_1 + P_2 + Q_1 + Q_2)}$$
$$O_2 = \overline{(P_2 + P_1 + Q_2 + Q_1)}$$
$$O_3 = \overline{(P_3 + P_0 + Q_3 + Q_0)}$$

$$(7.6)$$

As Figure 7.9 shows, only the programming part of nanoPLA is used to show the quadded XOR implementation on nanoPLA using the wrapping feature.

Firstly, the top inverting line buffers the inputs i_0 to i_3 and j_0 to j_3 twice to be used in the bottom ORed area. Secondly, the bottom plane generates the quadded OR forms and inverter of bottom plane inverts it. It also inverts the results of $S_{0..3}$ and $T_{0..3}$. They are then ORed again with the buffered $i_{0..3}$ and $j_{0..3}$ in a suitable manner for quadded logics. Finally, the inverter in the bottom plane inverts the result and generates $O_{0..3}$.

Although quadded logics have shown to be high-reliable [120], applying them on logics will result in a 4 times bigger size. However, in nanoPLA, and by using the wrapping feature, this area overhead is not the case anymore. One can define nanoPLAs with up to 100 nanowires [42].

In order to implement the XOR function, nanowires in programming part of nanoPLA and inverting/buffering lines are used and added together in a way to achieve the function. The entire quadded form is implemented in one nanoPLA by using more nanowires. Therefore, even for big circuits, the number of used

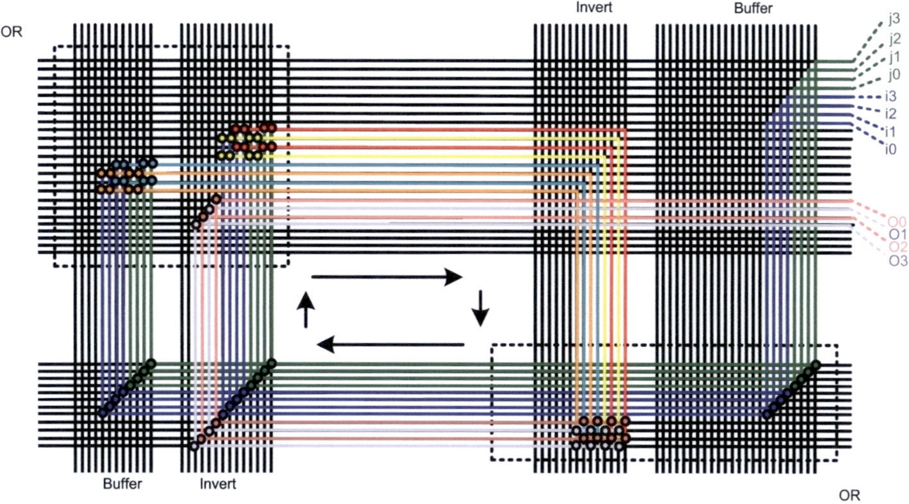

Figure 7.9: Quadded XOR on nanoPLA

nanoPLAs can be the same as the original design even for quadded form. Using quadded form, new nanowires in parallel are added to the original circuit wires.

7.7 Implementations

In this section, it is discussed how to implement the nanoPLA in a simulation environment to allow for fault injection and reliability evaluation.

A nanoPLA in VHDL is described by defining four internal wires: top and bottom buses and inverting and restoring buses in each every right and left parts of the nanoPLA. Inverting/restoring buses are connected to top (bottom) bus and then are ORed together in bottom (top) bus, achieving NOR logic, as shown in Figure 7.6.

The XOR mapping on nanoPLA in Figure 7.6 is used to make connections between corresponding buses. First, one has to buffer i and j in the top bus, then transfer them to the bottom bus. By inverting/buffering them in the left buses, one obtains them ORed together in the top bus, and they are then sent to the right plane to be inverted and form a NOR gate. Assuming TB, BB, RI, RB, LI and LB respectively as Top Bus, Bottom Bus, Right Inverter, Right Buffer, Left Inverter and Left Buffer, Figure 7.6 shows the performed computations to achieve a XOR gate.

7.7.1 Quadded Form

In a similar way to the XOR, quadded XOR form can be implemented as shown in Figure 7.9. In order to obtain the quadded gate, inputs are quadruplicated in the top bus and buffer them, and then are sent to the circle of nanoPLA.

After buffering inputs in Top-Right position, they are sent to the Down-Right part and are ORed in a proper manner to achieve quadded Logic properties. Then, in the Down-Right part they are inverted/buffered and are sent to the Top-Left part, where they are again ORed together to obtain the second level of the design in Figure 7.8(b). Finally, they are inverted in the Top-Left inverting lines and send them to the outputs.

7.7.2 TMR Form

In a similar fashion to the quadded form, the TMR form of an XOR is implemented to be compared to quadded form concerning defect tolerance. Every XOR logic gate (reconstructed in NOR gates) is triplicated, then a majority voter is inserted to compute the correct result. The voter is rewritten also in NOR form. The voter can be designed as shown in Equation 7.7. However, for TMR form of XOR, the function is triplicated on nanoPLA.

$$
\begin{aligned}
O = {} & \overline{(\overline{O_0} + \overline{O_1} + \overline{O_0})} \\
& + \overline{(\overline{O_0} + \overline{O_1} + O_2)} \\
& + \overline{(\overline{O_0} + O_1 + \overline{O_2})} \\
& + \overline{(O_0 + \overline{O_1} + \overline{O_2})}
\end{aligned}
\tag{7.7}
$$

TMR's implementation on nanoPLA is shown in Figure 7.10.

7.8 Experimental Results

To assess the reliability of the quadded form, the primary study focuses on the three different implementations of a simple XOR gate. The original XOR circuit, a TMR implementation with single voting, and the quadded XOR logic are tested. For each of these the experiments for all four combinations of the input vector are conducted and the results are measured. In order to influence the circuit, one, two or three faults are injected randomly in the design and compared the

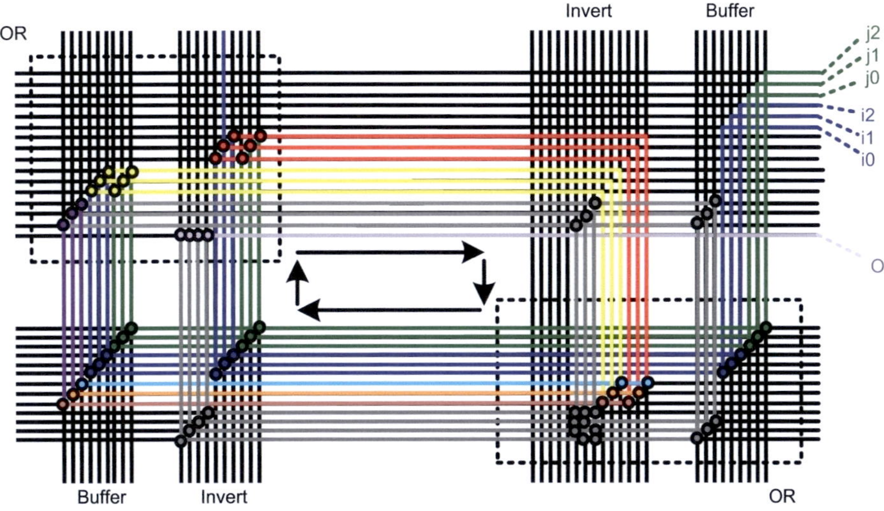

Figure 7.10: TMR implementation on nanoPLA. Gray wires show the voter implementation.

result of the faulty circuit to the expected result. In the framework fault injection in all nanowires of the nanoPLA is possible. Each of the faults corresponds to a typical defect in this kind of architecture. In this first experiment just one single nanoPLA is used. Defects and faults between different tiles will be considered in the future works.

The fault injector works as follows. For each run one, two, and three random faults are injected in the design by forcing the nanowires to one or zero and running a ModelSim simulation for this special case. This defect defines a stuck at one or zero on wires in the circuit. The fault injector checked the golden value of a wire, then flipped the value (stuck-at failure injection) and simulated the faulty circuit again. This is done 1000 times for each circuit and one to three faults. The number of 1000 runs per test case has been shown to be a good number as the results get stable and a significant amount of different fault combinations in different wires is tested. Moreover the setup guarantees that each single junction is tested once at least. Finally 1000 runs still give a reasonable runtime for the simulation.

Figure 7.11 shows the experimental results which include availability percentage of an XOR in original circuit, its TMR implementation and the quadded form for all different input vectors. 3 curves show experiments, which are done respectively for 1, 2, and 3 defects each time.

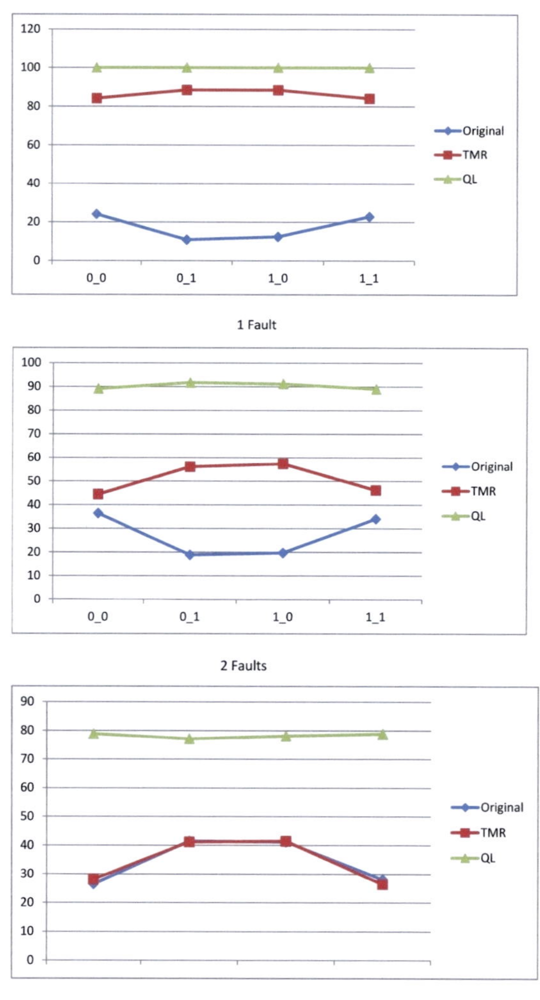

Figure 7.11: Availability percentage of an XOR in original circuit, TMRed and quadded form vs. different input distributions. Experiments are done for 1, 2, and 3 defects each time. In different distribution of inputs, different availability percentage was found, reported by the fault injector.

In the first curve, 1 defect is inserted in every simulation of the circuit. It means that, one nanowire is defective in every simulation and the result is checked for correctness. For quadded form, correction is done for all single defects as the curve shows. The availability is reported to be 100%. TMR achieves between 80% and 90% depending on the input value which is mainly caused by the voter as it is a single point of failure. The original implementation is correct in 10% to 25% depending on the input vectors.

In the second curve, experiments are repeated for 2 defect injections in every simulation. The correctness of the original circuit increases to 20% to 40% caused by two compensating faults. The TMR reliability decreases because two copies can be affected at the same time. The quadded logic is still able to correct up to 90% of the defects.

In the third curve, quadded form still gives a correct result for up to 80% for the test cases. Interestingly TMR gives almost the same result as the simple XOR implementation.

7.9 Discussion

By using different input vectors, different availability percentages were found, reported by the fault injector. It is due to the partial fault correctness feature of XOR logic gate. For example, when two input vectors have a different value, the output value is always one. Now, if both of inputs are faulty and change the value (one to zero and vice versa), the output will still remain the correct value. In the results can also be seen that different input vectors result in different availabilities of the circuit in the existence of defects.

As Figure 7.11 shows, quadded logics are able to tolerate against single defects in all cases based on the experiments and also the proof in [120]. As shown, TMR is unreliable as the number of defects increases. As already noted, a voter in TMR is a single point of failure which hugely effect the reliability.

In large scales, quadded logics are able to make designs high-reliable. As mentioned before, the quadded NOR logics can be used in nanoPLA to get the benefits of quadded NOR logics reliability on nanoPLAs. NOR logic gates can be used to implement more complex systems and quadded logics are more reliable when there are more levels between designed NOR gates.

In comparison to the conventional logic mapping on nanoPLA [40], the method allows for faster mapping. The process of test and characterization of nanoPLA is eliminated and quadded method tolerates all single defects and most of multiple defects in design.

8 Conclusion and Future work

As described in the beginning, the work presented here aims at designing an efficient approach for high reliability of FPGA based systems under high radiation.

The possibility to integrate a complex Systems-on-Chip (SoC) on a single FPGA device, combining speedup of hardware with flexibility of software, the opportunity of practically unlimited reconfiguration in SRAM based FPGAs as well as drastically sinking prices, have made FPGAs attractive for a wide variety of application domains.

In special application domains such as space applications, FPGAs are inherently flexible to meet multiple requirements and offer significant performance and cost saving for such critical applications. Therefore the popularity to use such devices in space applications has increased in recent years. Especially SRAM-based FPGAs which embed megabits of RAMs and plenty of configurable logic and routing resources, makes it possible to implement circuits composed of millions of gates which are tailored to the application.

The main motivation for studying fault tolerance techniques for FPGAs, is the lack of robustness in modern technologies due to the following reasons: First, state of the art nano scale technologies in general and SRAM based FPGAs in particular have been shown to be susceptible to radiation in the aforementioned environment. In fact, these effects can cause transient or permanent bit flipping on SRAM cells and respectively change the function of logic elements within FPGAs.

In addition, as technology size decreases below nanometers, several major single event effect challenges related to nuclear and space radiation arise. Experiments show a huge increase in SEU sensitivity for each new device generation using shrinking technology. In addition, MBUs which are the result of SEUs are more likely to provoke more than one bit upset. The width of the area which is affected by MBUs is likely to have an influence on fault tolerance methodologies.

Traditional methods such as triple modular redundancy approaches have been proved to be ineffective in mitigating against SEEs with relatively high rates. Along with decreasing technology scales this causes the coarse grain TMR to face some challenges.

In addition with multiple SETs traditional coarse grain TMR approaches might be unable to detect and recover from an error state. This can be improved when redundancy is used on smaller portions of the design, hence going to a more fine grain view.

In this work, methodologies are studied and proposed which focus on solving the aforementioned challenges by changing the granularity of fault tolerance to very fine grain levels. The work is composed of two improvements on conventional redundancy. The basic principle that has been followed, is to change the granularity of fault tolerance in FPGAs.

In the first step, suitable granular redundancy for FPGAs is examined and the suitable granularity level is studied. The homogeneous nature of FPGAs allows to find a common level where a methodology can be applied without being focused to a certain design. Coupling the redundancy to FPGA primitives allows to apply new considerations in using fault tolerance methodologies.

This way the FGTMR technique is introduced in this work as an improvement to TMR. It allows to recover from failures on local parts of the architecture. This comes to the price that every TMRed fine grain part needs a voter to select the correct output value which results in a longer critical path. Therefore in the second step of this contribution the focus is on this challenge. A possible solution are methodologies which don't need voting and are still applicable to fine grains.

Therefore a methodology is discussed which uses inherent voting in fine grain parts of FPGAs by wiring different copies of a grain in a special manner and using them to vote for correct values. The idea is based on quadded logic and is enhanced in order to work on modern reconfigurable architectures. The new method is named QFDR. It applies two different rules to the boolean functions (LUTs in FPGAs) and realizes inherent voting by connecting the LUTs in a special way.

Using these approaches, a tool flow is introduced which has been developed in this work and can be easily integrated in standard synthesis tools for FPGAs. This tool flow allows automatically applying FGTMR or QFDR to any design after the mapping process. As case study, the Xilinx ISE tool flow is used to integrate fault tolerance methodologies on the Xilinx Virtex-5 family. The tool flow modifies designs in lower levels of the synthesis flow and applies QFDR or FGTMR to them. Experiments are done on different designs and results on area overhead, power usage, and fault tolerance are presented. The trade-off between high reliability and resource consumption based on the aforementioned methodologies is eventually discussed.

Eventually both FGTMR and QFDR show an improvement in reliability. Of course this comes to the price of additional area consumption. However this overhead can be significantly reduced as soon as slight changes to the FPGA

LUT architecture are considered. Therefore an improved slice presentation is introduced. In addition this allows to simplify the methodology of automatically applying QFDR on future FPGAs.

Finally, in a longer perspective, nano architectures are studied in this work. These architectures are likely to replace the CMOS technology because as soon as the end of the scaling road map is reached. Therefore trends on new nano architectures and defect tolerance as their main challenge are discussed in this contribution. These architectures show some similarities to current reconfigurable architectures. Especially high failure and error rates are expected for these nanoPLAs due to the fabrication process. Hence fault tolerance methodologies are absolutely necessary. In this work is shown, that quadded logic is applicable for nanoPLAs, provides good results and even can simplify the characterization process.

As in any methodology there are some limitations and some challenges remain. The current technique which is used in QFDR, assumes that in the case of flipflop with enable, voter is inserted to provide the correct output value to be written back to the flipflop. Although this implementation recovers flipflops in case of SEUs, it adds an overhead of voters to the design. One of the major next steps of this work can be extending QFDR in order to cope with SEUs in flipflops which use enable.

Also the area overhead of the proposed methodologies is a challenge. It can be improved by using mechanisms which optimize the placement and routing issues in a synthesis flow as has been shown in the Spartan-3 synthesis Virtex-5 P&R experiment. In principle, high-level synthesis can be done for 3-LUT slices or less and be optimized for a special number of inputs in LUTs in order to apply QFDR without needing for LUT decomposition. This way a better optimization is feasible.

Another challenge are MBUs. In this work, a methodology was presented which is based on placement modifications. This can only be seen as a principle and needs some refinement.

In conclusion the results of this work show, that changing the granularity level of redundancy on SRAM based FPGAs brings some benefits whenever high error rates are being considered. First of all a higher fault tolerance compared to coarse grain approaches is achieved by applying the new methodologies that have been developed in this work. Secondly using FPGA primitives for tolerance insertion allows to fully automatically integrate redundancy to any FPGA design in principle. Eventually this work demonstrates how quadded logic even can be applied to future nanotube based architectures beyond the CMOS roadmap.

List of Figures

List of Tables

Abbreviations

ASIC	Application Specific Integrated Circuit
BRAM	Block Random Accessible Memory
CAD	Computer Aided Design
CCD	Charge Coupled Devices
CFG	Configuration
CGTMR	Coarse grain TMR
CLB	Configurable Logic Block
CMOS	Complementary Metal Oxide Semiconductor
CRC	Cyclic Redundancy Code
DFF	D-Flip Flop
DSP	Digital Signal Processor
DTMR	Distributed TMR
EEPROM	Electrically EPROm
EPROM	Earasable PROM
FF	Flip Flop
FGTMR	Fine Grain TMR
FPGA	Field Programmable Gate Array
GEO	Geoscnchronous
GPS	Flobal Positioning System
GTMR	Global TMR
HDL	Hardware Description Language
IC	Integrated Circuit
IOB	Input Output Block
IP	Intellectual Property
JEDEC	Joint Electron Devices Engineering Council
KCPSM	Ken Chapmans Programmable State Machine
LEO	Low Earth Orbit
LET	Linear Energy Transfer
LTMR	Local TMR
LUT	Look Up Table
MBU	Multiple Bit Upset
MUX	Multiplexer
NASA	National Aeronautics and Space Administration
NCD	Natural Circuit Description

NW	Nano Wire
PAL	Programmable Array Logic
PIP	Programmable Interconnection Point
PLA	Programmable Logic Array
PLD	Programmable Logic Device
PROM	Programmable Read-only Memory
QFDR	Quadded Force Decide Redundancy
QL	Quadded Logic
ROM	Read Only Memory
RoRA	Reliability Oriented place and Route Algorithm
RTL	Register Transfer Level
SE	Single Event
SEB	Single Event Burnout
SEE	Single Event Effect
SEGR	Single Event Gate Rupture
SEL	Single Event Latchup
SET	Single Event Transient
SEU	Single Event Upset
SOC	System On Chip
SRAM	Static Random Accessible Memory
TCL	Tool Command Language
TID	Total Ionizing Dose
TMR	Tripple Modular Redundancy
VHDL	Very High Speed Description Language
XDL	Xilinx Design Language
XOR	Exclusive OR
XTMR	Xilinx TMR

Bibliography

[1] *http://beta.ivc.no/blog/2011/03/30/logic-devices/.* In *FPGA, CPLD, and EPP Solution from Xilinx.*

[2] *http://www.actel.com/.* In *Low Power FPGAs and Mixed Signal FPGAs.*

[3] *http://www.actel.com/documents/PA3DS.pdf.* In *ProASIC3 Flash Family FPGAs.*

[4] *http://www.altera.com/.* In *FPGA, CPLD and ASIC from Altera.*

[5] *http://www.altera.com/literature/ds/classic.pdf.* In *Classic EPLD family data sheet A-DS-CLASSIC-05.*

[6] *http://www.latticesemi.com/.* In *FPGA and CPLD solution from Lattice Semiconductor.*

[7] *http://www.model.com/.* In *ModelSim - Advanced Simulation and Debugging.*

[8] *http://www.quicklogic.com/images/polarproDS.pdf.* In *QuickLogic PolarPro Data Sheet (Rev H).*

[9] *http://www.xilinx.com/.* In *FPGA, CPLD, and EPP Solution from Xilinx.*

[10] *NASA/GSFC Radiation Effects and Analysis home page.*

[11] *The Synthesis of Two-Terminal Switching Circuits.* Bell System Technical Journal Volume XXVIII Number 1, November 1949.

[12] *International Technology Roadmap for Semiconducters(ITRS),* 2005.

[13] *Single-Event Upset Mitigation for Xilinx FPGA Block Memories.* In *www.xilinx.con,* vol. XAPP962, March 2008.

[14] ABRAHAM, J.: *A Combinatorial Solution to the Reliability of Interwoven Redundant Logic Networks.* Computers, IEEE Transactions on, C-24(5):578 – 584, may 1975.

[15] ABRAHAM, J. A.: *Reliability analysis of digital systems protected by massive redundancy.* PhD thesis, Stanford University, 1974.

[16] ALDERIGHI, M., F. CASINI, S. D'ANGELO, M. MANCINI, D. MERODIO CODINACHS, S. PASTORE, G. SORRENTI, L. STERPONE, R. WEIGAND and M. VIOLANTE: *Robustness analysis of soft error accumulation in SRAM-FPGAs using FLIPPER and STAR/RoRA.* In *Radiation and Its Effects on Com-*

ponents and Systems (RADECS), 2008 European Conference on, pp. 157 –161, sept. 2008.

[17] ALLEN, J. A. V.: *The geomagnetically-trapped corpuscular radiation*. Dept. of Physics, State University of Iowa, 1956.

[18] ANDERSON, S. A.: *Failure Estimation for Partial TMR Mitigated Designs in a Virtex-4 FPGA*, 2010.

[19] AZAMBUJA, J., F. SOUSA, L. ROSA and F. KASTENSMIDT: *Evaluating large grain TMR and selective partial reconfiguration for soft error mitigation in SRAM-based FPGAs*. In *On-Line Testing Symposium, 2009. IOLTS 2009. 15th IEEE International*, pp. 101 –106, june 2009.

[20] B. P. WONG, A. MITTAL, Y. C. and G. STARR: *Nano-CMOS Circuit and Physical Design*. John Wiley and Sons, Inc., 2005.

[21] BAO LIU, E. B. Z. W.: *Nanoelectronic Design Based ona CNT Nano-Architecture*. Publischer:InTech, February 2010, ISBN 978-953-307-049-0, 2010.

[22] BECKHOFF, C., D. KOCH and J. TORRESEN: *The Xilinx Design Language (XDL): Tutorial and use cases*. In *Reconfigurable Communication-centric Systems-on-Chip (ReCoSoC), 2011 6th International Workshop on*, pp. 1 –8, june 2011.

[23] BERG, M.: *https://nepp.nasa.gov/mafa/talks /MAFA0734Berg.pdf*.

[24] BERG, M., C. POIVEY, D. PETRICK, D. ESPINOSA, A. LESEA, K. LABEL, M. FRIENDLICH, H. KIM and A. PHAN: *Effectiveness of Internal Versus External SEU Scrubbing Mitigation Strategies in a Xilinx FPGA: Design, Test, and Analysis*. Nuclear Science, IEEE Transactions on, 55(4):2259 –2266, aug. 2008.

[25] BIRKNER, J., A. CHAN, H. CHUA, A. CHAO, K. GORDON, B. KLEINMAN, P. KOLZE and R. WONG: *A very-high-speed field-programmable gate array using metal-to-metal antifuse programmable elements*. Microelectronics Journal, 23(7):561 – 568, 1992.

[26] BOBDA, C.: *Introduction to Reconfigurable Computing*. Springer, 2007.

[27] BOZHUYUK, O.: *Placement modification of Virtex5 FPGAs to tolerate against MBUs*, 2012.

[28] BRENDAN BRIDGFORD, C. C. and C. W. TSENG: *Single-Event Upset Mitigation Selection Guide*. In *www.xilinx.con*, vol. XAPP987, March 2008.

[29] BUTTS, M., A. DEHON and S. GOLDSTEIN: *Molecular electronics: devices, systems and tools for gigagate, gigabit chips*. In *Computer Aided Design, 2002. ICCAD 2002. IEEE/ACM International Conference on*, pp. 433 – 440, nov. 2002.

[30] CAMPOSANO, R.: *From behavior to structure: high-level synthesis*. Design

Test of Computers, IEEE, 7(5):8 –19, oct. 1990.

[31] CARMICHAEL, C. CAFFREY, M. and S. A.: *Correcting Single Event Upset through Virtex Partial Reconfiguration.* In *Xilinx Application Notes, XAPP216,* 2000.

[32] C.CARMICHAEL: *Correcting Single Event Upsets through Virtex Partial Reconfiguration.* In *www.xilinx.con,* vol. Xilinx Application Notes, June 2000.

[33] C.CARMICHAEL: *Triple Module Redundancy Design Techniques for Virtex FPGAs.* In *www.xilinx.con,* vol. XAPP197, November 2001.

[34] CHEN, J., S. ELTOUKHY, S. YEN, R. WANG, F. ISSAQ, G. BAKKER, J. YEH, E. POON, D. LIU and E. HAMDY: *A modular 0.8 mu m technology for high performance dielectric antifuse field programmable gate arrays.* In *VLSI Technology, Systems, and Applications, 1993. Proceedings of Technical Papers. 1993 International Symposium on,* pp. 160 –164, 1993.

[35] CHEN, Y., STEWART, R. DUNCAN, W. STANLEY, JEPPESEN, O. JAN, K. NIELSEN, STODDART, L. OLYNICK DEIRDRE and E. ANDERSON: *Nanoscale molecular-switch devices fabricated by imprint lithography.* Applied Physics Letters, 82(Issue 10, id. 1610), 2003.

[36] CHIANG, S., R. FOROUHI, W. CHEN, F. HAWLEY, D. MCCOLLUM, E. HAMDY and C. HU: *Antifuse structure comparison for field programmable gate arrays.* In *Electron Devices Meeting, 1992. IEDM '92. Technical Digest., International,* pp. 611 –614, dec 1992.

[37] COPEN GOLDSTEIN, S. and M. BUDIU: *NanoFabrics: spatial computing using molecular electronics.* In *Computer Architecture, 2001. Proceedings. 28th Annual International Symposium on,* pp. 178 –189, 2001.

[38] D. MCMURTREY, K. MORGAN, B. P. and M.WIRTHLIN: *D. McMurtrey, K. Morgan, B. Pratt, and M. Wirthlin, Estimating TMR Reliability on FPGAs Using Markov Models.* Available: http://dspace.byu.edu, 2006.

[39] DEHON, A.: *Array-based architecture for FET-based, nanoscale electronics.* Nanotechnology, IEEE Transactions on, 2(1):23 – 32, mar 2003.

[40] DEHON, A.: *Nanowire-based programmable architectures.* J. Emerg. Technol. Comput. Syst., 1:109–162, July 2005.

[41] DEHON, A., P. LINCOLN, J. E. SAVAGE and L. FELLOW: *Stochastic assembly of sublithographic nanoscale interfaces.* IEEE Transactions on Nanotechnology, 2:165–174, 2003.

[42] DEHON, A. and M. J. WILSON: *Nanowire-based sublithographic programmable logic arrays.* In *Proceedings of the 2004 ACM/SIGDA 12th international symposium on Field programmable gate arrays, FPGA '04,* pp. 123–132, New York, NY, USA, 2004. ACM.

[43] DING, Y.: *Study of Radiation-Tolerant Integrated Circuits for Space Applications*, 2010.

[44] DINGWALL, A. and R. STRICKER: *High density COS/MOS 1024-bit static random access memory*. Solid-State Circuits, IEEE Journal of, 10(4):197 – 200, aug 1975.

[45] DOUCIN, B., T. CARRIERE, C. POIVEY, P. GARNIER, J. BEAUCOUR and Y. PATIN: *Model of single event upsets induced by space protons in electronic devices*. In *Radiation and its Effects on Components and Systems, 1995. RADECS 95., Third European Conference on*, pp. 402 –408, sep 1995.

[46] DUZELLIER, S.: *Component characterization and testing: Single event*. Radecs short course, 2003.

[47] FROHMAN-BENTCHKOWSKY, D.: *Floating Gate Transistor and method for charging and discharging same*, June 1970.

[48] FULLER, E., M. CAFFREY, P. BLAIN, C. CARMICHAEL, N. KHALSA and A. SALAZAR: *Radiation test results of the Virtex FPGA and ZBT SRAM for Space Based Reconfigurable Computing*. In *Proceeding of the Military and Aerospace Programmable Logic Devices International Conference(MAPLD)*, Laurel, MD, Sept. 1999.

[49] FULLER, E., M. CAFFREY, A. SALAZAR, C. CARMICHAEL and J. FABULA: *Radiation Testing Update, SEU Mitigation, and Availability Analysis of the Virtex FPGA for Space Re-configurable Computing, presented at the IEEE Nuclear and Space Radiation Effects Conference*. In *in Proc. International Conference on Military and Aerospace Programmable Logic Devices*, 2000.

[50] GHOSH, S.: *XDL Based Hard Macro Generator*, 2011.

[51] GOERGE PERLEGOS, W. S. J.: *Electrically Alterable Read Mostly Memory*, Feb 1979.

[52] GOJMAN, B., H. MANEM, G. ROSE and A. DEHON: *Inversion schemes for sublithographic programmable logic arrays*. Computers Digital Techniques, IET, 3(6):625 –642, november 2009.

[53] GOLKE, K.: *Determination of funnel length from cross section versus LET measurements*. Nuclear Science, IEEE Transactions on, 40(6):1910 –1917, dec 1993.

[54] GORDON, K. and R. WONG: *Conducting filament of the programmed metal electrode amorphous silicon antifuse*. In *Electron Devices Meeting, 1993. IEDM '93. Technical Digest., International*, pp. 27 –30, dec 1993.

[55] GRAHAM, P., M. CAFFREY, D. JOHNSON, N. ROLLINS and M. WIRTHLIN: *SEU mitigation for half-latches in Xilinx Virtex FPGAs*. Nuclear Science, IEEE Transactions on, 50(6):2139 – 2146, dec. 2003.

[56] GREENE, J., E. HAMDY and S. BEAL: *Antifuse field programmable gate arrays.* Proceedings of the IEEE, 81(7):1042 –1056, jul 1993.

[57] GUTERMAN, D., I. RIMAWI, T.-L. CHIU, R. HALVORSON and D. MCELROY: *An electrically alterable nonvolatile memory cell using a floating-gate structure.* Electron Devices, IEEE Transactions on, 26(4):576 – 586, apr 1979.

[58] HAMDY, E., J. MCCOLLUM, S.-O. CHEN, S. CHIANG, S. ELTOUKHY, J. CHANG, T. SPEERS and A. MOHSEN: *Dielectric based antifuse for logic and memory ICs.* In *Electron Devices Meeting, 1988. IEDM '88. Technical Digest., International,* pp. 786 –789, 1988.

[59] HAUNG, Y.: *Logic Gates and Computation from Assembled Nanowire Building Blocks.* Science, 2002.

[60] HEATH, J. R., P. J. KUEKES, G. S. SNIDER and R. S. WILLIAMS: *A defecttolerant computer architecture: Opportunities for nanotechnology.* Science, 280:1716–1721, 1998.

[61] HORST, J. G. VAN DER: *Radiation Tolerant Implementation of a soft-core processor for space applications.* PhD thesis, University of Stellenbosch, Souf Africa, 2007.

[62] HUANG, J., M. B. TAHOORI and F. LOMBARDI: *On the Defect Tolerance of Nano-Scale Two-Dimensional Crossbars.* Defect and Fault-Tolerance in VLSI Systems, IEEE International Symposium on, 0:96–104, 2004.

[63] INC., X. In *UG190: Virtex-5 FPGA User Guide, 2009. v5.2, November,* November 2009.

[64] INC., X. In *UG632: Xilinx Plan Ahead User Guide, 2009. v 11.4, December,* December 2009.

[65] INC., X. In *UG129: PicoBlaze 8-bit Embedded Microcontroller User Guide,* January 2010.

[66] INC., X.: *Spartan-3 FPGA Family Data Sheet.* In *Xilinx Datasheet DS009,* Dec. 2009.

[67] INC., X.: *Virtex 2.5 V Field Programmable Gate Arrays.* In *Xilinx Datasheet DS003, v2.4,* Oct. 2000.

[68] ITURBE, X., M. AZKARATE, I. MARTINEZ, J. PEREZ and A. ASTARLOA: *A novel SEU, MBU and SHE handling strategy for Xilinx Virtex-4 FPGAs.* In *Field Programmable Logic and Applications, 2009. FPL 2009. International Conference on,* pp. 569 –573, 31 2009-sept. 2 2009.

[69] J., N. R., R. P. A., K. K., S. W. J., M. P. T. and H. W.: *Radiation Characterization of a Hardened 0.22 956;m Anti-Fuse Field Programmable Gate Array.* Nuclear Science, IEEE Transactions on, 53(6):3525 –3531, dec. 2006.

[70] JENSEN, P. A.: *Quadded NOR Logic.* Reliability, IEEE Transactions on, R-

12(3):22 –31, sept 1963.

[71] J.GOLDBERG, H.S.STONE, A.: *TECHNIQUES FOR THE REALIZA-
TION OF ULTRARELIABLE SPACEBORNE COMPUTERS*. In *Stanford Research Institute*, December 1968.

[72] JULIATO, M.: *Fault Tolerant Cryptographic Primitives for Space Applications*.
PhD thesis, University of Waterloo, Canada, 2011.

[73] KASTENSMIDT, F., L. STERPONE, L. CARRO and M. REORDA: *On the optimal design of triple modular redundancy logic for SRAM-based FPGAs*. In *Design, Automation and Test in Europe, 2005. Proceedings*, pp. 1290 – 1295 Vol. 2,
march 2005.

[74] KEAN, T.: *Secure Configuration of Field Programmable Gate Arrays*. In *In International Conference on Field-Programmable Logic and Applications 2001 (FPL
2001*, pp. 142–151. Springer-Verlag, 2001.

[75] KOOMEY, J., S. BERARD, M. SANCHEZ and H. WONG: *Implications of His-
torical Trends in the Electrical Efficiency of Computing*. Annals of the History
of Computing, IEEE, 33(3):46 –54, march 2011.

[76] KUON, I. and J. ROSE: *Measuring the Gap Between FPGAs and ASICs*.
Computer-Aided Design of Integrated Circuits and Systems, IEEE Transactions on, 26(2):203 –215, feb. 2007.

[77] KUON, I., R. TESSIER and J. ROSE: *FPGA Architecture: Survey and Challenges*.

[78] KUWAHARA, T.: *FPGA-based Reconfigurable On-board Computing Systems for
Space Applications*. PhD thesis, PhD work at Universiteat Stuttgart, 2010.

[79] LEVENTIS, P., B. VEST, M. HUTTON and D. LEWIS: *MAX II: A low-cost,
high-performance LUT-based CPLD*. In *Custom Integrated Circuits Conference,
2004. Proceedings of the IEEE 2004*, pp. 443 – 446, oct. 2004.

[80] LIMA, F., L. CARRO and R. REIS: *Designing fault tolerant systems into
SRAM-based FPGAs*. In *Design Automation Conference, 2003. Proceedings*, pp.
650 – 655, June 2003.

[81] LOVELESS, T. D.: *A RADIATION-HARDENED-BY-DESIGN CHARGE
PUMP FOR PHASE-LOCKED LOOP CIRCUITS*, 2007.

[82] LYONS, R. E. and W. VANDERKULK: *The Use of Triple-Modular Redundancy
to Improve Computer Reliability*. IBM Journal of Research and Development,
6(2):200 –209, april 1962.

[83] MARKOV, A.: *Extension of the Limit Theorems of Probability Theory to a Sum
of Variables Connected in a Chain*. In *In Dynamic Probabilistic Systems*, vol.
Volume I: Markov Models, 1971.

[84] MORRIS, R. and C. FOSTER: *Cross-Section Estimation in the Presence of Un-*

certain Dosimetry. Nuclear Science, IEEE Transactions on, 57(6):3528 –3536, dec. 2010.

[85] MUCHA, I.: *Ultra Low Voltage Class AB Switched Current Memory Cells Based on Floating Gate Transistors.* Analog Integr. Circuits Signal Process., 20:43–62, July 1999.

[86] NAEIMI, H.: *Reliable integration of terascale systems with nanoscale devices.* PhD thesis, California Institute of Technology, 2007.

[87] NAEIMI, H. and A. DEHON: *A greedy algorithm for tolerating defective crosspoints in nanoPLA design.* In *In FPT*, 2004.

[88] NAVABI, Z.: *Embedded Core Design with FPGAs.* McGraw-Hill, August 2006.

[89] NICOLAIDIS, M.: *Time redundancy based soft-error tolerance to rescue nanometer technologies.* In *VLSI Test Symposium, 1999. Proceedings. 17th IEEE*, pp. 86 –94, 1999.

[90] NIKNAHAD, M., O. SANDER and J. BECKER: *QFDR-an integration of Quadded Logic for modern FPGAs to tolerate high radiation effect rates.* September 2011.

[91] NIKNAHAD, M., O. SANDER and J. BECKER: *A study on fine granular fault tolerance methodologies for FPGAs.* In *Reconfigurable Communication-centric Systems-on-Chip (ReCoSoC), 2011 6th International Workshop on*, pp. 1 –5, june 2011.

[92] NIKNAHAD, M., O. SANDER and J. BECKER: *FGTMR - Fine Grain Redundancy Method for Reconfigurable Architectures under high Failure Rates.* NAS-NIT 2011.

[93] OPENCORES. In *http://opencores.org/projects*.

[94] OSTLER, P., M. CAFFREY, D. GIBELYOU, P. GRAHAM, K. MORGAN, B. PRATT, H. QUINN and M. WIRTHLIN: *SRAM FPGA Reliability Analysis for Harsh Radiation Environments.* Nuclear Science, IEEE Transactions on, 56(6):3519 –3526, dec. 2009.

[95] PAULSSON, K.: *Power Optimized Design of FPGA-based Self-Adaptive Systems.* PhD thesis, Universitaet Karlsruhe (TH), 2009.

[96] PAULSSON, K., M. HUBNER, M. JUNG and J. BECKER: *Methods for run-time failure recognition and recovery in dynamic and partial reconfigurable systems based on Xilinx Virtex-II Pro FPGAs.* In *Emerging VLSI Technologies and Architectures, 2006. IEEE Computer Society Annual Symposium on*, vol. 00, p. 6 pp., march 2006.

[97] PETERSEN, E.: *Cross section measurements and upset rate calculations.* Nuclear Science, IEEE Transactions on, 43(6):2805 –2813, dec 1996.

[98] PRATT, B., M. CAFFREY, J. CARROLL, P. GRAHAM, K. MORGAN and

M. WIRTHLIN: *Fine-grain SEU mitigation for FPGAs using Partial TMR*. In *Radiation and Its Effects on Components and Systems, 2007. RADECS 2007. 9th European Conference on*, pp. 1 –8, sept. 2007.

[99] PRATT, B. H.: *Analysis and Mitigation of SEU-induced Noise in FPGA-based DSP Systems*. PhD thesis, Brigham Young University, 2011.

[100] QUINN, H., P. GRAHAM, J. KRONE, M. CAFFREY and S. REZGUI: *Radiation-induced multi-bit upsets in SRAM-based FPGAs*. Nuclear Science, IEEE Transactions on, 52(6):2455 – 2461, dec. 2005.

[101] QUINN, H., K. MORGAN, P. GRAHAM, J. KRONE and M. CAFFREY: *Static Proton and Heavy Ion Testing of the Xilinx Virtex-5 Device*. In *Radiation Effects Data Workshop, 2007 IEEE*, vol. 0, pp. 177 –184, july 2007.

[102] RAI-CHOUDHURY, P.: *Handbook of microlithography, micromachining, and microfabrication*, vol. 1. SPIE press, 1997.

[103] RAPIDSMITH: *http://rapidsmith.sourceforge.net/doku.php?id=home*.

[104] REN, A. and Q. YIN: *FPGA implementation of a W-CDMA system based on IP functions*. In *Proceedings of the 2005 WSEAS international conference on Dynamical systems and control*, CONTROL'05, pp. 320–324, Stevens Point, Wisconsin, USA, 2005. World Scientific and Engineering Academy and Society (WSEAS).

[105] RHOD, E. L. and L. CARRO: *An efficient test and characterization approach for nanowire-based architectures*. In *Proceedings of the 21st annual symposium on Integrated circuits and system design*, SBCCI '08, pp. 34–39, New York, NY, USA, 2008. ACM.

[106] RICHARD H. MAURER, MARTIN E. FRAEMAN, M. N. M. and D. R. ROTH: *Harsh Environments: Space Radiation Environment, Effects, and Mitigation*. John Hopkins Apl Technical Digest, 28(1), 2008.

[107] ROOSTA, R.: *A Comparison of Radiation-Hard and Radiation-Tolerant FPGAs for Space Applications*. NASA Electronic Parts and Packaging Program, NASA and Jet Propulsion Laboratory, December 2004.

[108] RUHL, K.: *An introduction to Space Weather*. Computer Sceintists, March 2010.

[109] S., R., L. P. and S. R.: *SEE Characterization of the New RTAX-DSP (RTAX-D) Antifuse-Based FPGA*. Nuclear Science, IEEE Transactions on, 57(6):3537 –3546, dec. 2010.

[110] SCHALLER, R. R.: *Moore's law: past, present, and future*. IEEE Spectr., 34:52–59, June 1997.

[111] SHIH, C.-C., R. LAMBERTSON, F. HAWLEY, F. ISSAW, J. MCCOLLUM, E. HAMDY, H. SAKURAI, H. YUASA, H. HONDA, T. YAMAOKA, T. WADA

and C. HU: *Characterization and modeling of a highly reliable metal-to-metal antifuse for high-performance and high-density field-programmable gate arrays.* In *Reliability Physics Symposium, 1997. 35th Annual Proceedings., IEEE International*, pp. 25 –33, apr 1997.

[112] SHIVAKUMAR, P., M. KISTLER, S. KECKLER, D. BURGER and L. ALVISI: *Modeling the effect of technology trends on the soft error rate of combinational logic.* In *Dependable Systems and Networks, 2002. DSN 2002. Proceedings. International Conference on*, pp. 389 – 398, 2002.

[113] SINNOTT, R., A. ASENOV, D. BERRY, D. CUMMING, S. FURBER, C. MILLAR, A. MURRAY, S. PICKLES, S. ROY, A. TYRRELL and M. ZWOLINKSI: *Meeting the Design Challenges of nanoCMOS Electronics: An Introduction to an EPSRC.* In *Pilot Project Proc. Of the Fifth UK e-Science All Hands Meeting (NeSC ISBN*, pp. 0–9553988, 2006.

[114] SKOROBOGATOV, S.: *Low temperature data remanence in static RAM.* Techn. Rep. UCAM-CL-TR-536, University of Cambridge, Computer Laboratory, June 2002.

[115] SMITH, M. D., D. JOHN, P. SHEWCHUK, M. D. SMITH, C. CHAIRMAN, D. WILLIAM and G. SULLIVAN: *ESTIMATION OF FUTURE MANUFACTURING COSTS FOR NANOELECTRONICS TECHNOLOGY*, 1996.

[116] STERPONE, L., M. REORDA and M. VIOLANTE: *RoRA: a reliability-oriented place and route algorithm for SRAM-based FPGAs.* In *Research in Microelectronics and Electronics, 2005 PhD*, vol. 1, pp. 173 – 176 vol.1, July 2005.

[117] STERPONE, L., M. VIOLANTE and S. REZGUI: *An Analysis Based on Fault Injection of Hardening Techniques for SRAM-Based FPGAs.* Nuclear Science, IEEE Transactions on, 53(4):2054 –2059, aug. 2006.

[118] TAHOORI, M.: *A mapping algorithm for defect-tolerance of reconfigurable nano-architectures.* Computer-Aided Design, International Conference on, 0:668–672, 2005.

[119] TEOSTE, R.: *Digital Circuit Redundancy.* Reliability, IEEE Transactions on, R-13(2):42 –61, june 1964.

[120] TRYON, J. G.: *Redundant Logic Circuitry.* to Bell Telephone Labs, Inc., July 1958.

[121] TRYON, J. G.: *Quadded logic.* Redundancy Techniques for Computing Systems by R. H. Wilcox and W. C.Mann Eds., Spartan Books, 1962.

[122] VASUDEVAN, V.: *Investigation of FPGA based architectures for safety critical applications.* PhD thesis, The university of Queensland, 2006.

[123] VIAL, J., A. BOSIO, P. GIRARD, C. LANDRAULT, S. PRAVOSSOUDOVITCH and A. VIRAZEL: *Using TMR Architectures for Yield Improvement.* In *De-*

fect and Fault Tolerance of VLSI Systems, 2008. DFTVS '08. IEEE International Symposium on, pp. 7 –15, oct. 2008.

[124] W., W.: *A statistical distribution function of wide applicability.* J. Appl. Mech.-Trans. ASME, 1951.

[125] YUI, C., G. SWIFT, C. CARMICHAEL, R. KOGA and J. GEORGE: *SEU mitigation testing of Xilinx Virtex II FPGAs.* In *Radiation Effects Data Workshop, 2003. IEEE,* pp. 92 – 97, july 2003.

Publications

Conference & Journalpapers

[NB08a]   M. Niknahad and J. Becker. Depth first traversal algorithm for efficient build-in self-test in nano fabrics. In *Reconfigurable Communication-centric Systems-on-Chip (ReCoSoC), 2008 3rd International Workshop on*, pages 1 –5, july 2008.

[NB08b]   M. Niknahad and J. Becker. Dynamic reconfiguration of nano architectures using application independent fault detection. In *In AMWAS 2008 School and Workshop*, October 2008.

[NHB09]   M. Niknahad, M. Hubner, and J. Becker. Method for improving performance in online routing of reconfigurable nano architectures. In *SOC Conference, 2009. SOCC 2009. IEEE International*, pages 65 –68, sept. 2009.

[NHB10]   M. Niknahad, M. Huebner, and J. Becker. Reliability analysis and improvement in nano scale design. In *VLSI (ISVLSI), 2010 IEEE Computer Society Annual Symposium on*, pages 299 –303, july 2010.

[NSB11a]   M. Niknahad, O. Sander, and J. Becker. Fgtmr - fine grain redundancy method for reconfigurable architectures under high failure rates. In *Nano, Information Technology and Reliability (NASNIT), 2011 15th North-East Asia Symposium on*, pages 186 –191, oct. 2011.

[NSB11b]   M. Niknahad, O. Sander, and J. Becker. Qfdr-an integration of quadded logic for modern fpgas to tolerate high radiation effect rates. In *Radiation and Its Effects on Components and Systems (RADECS), 2011 12th European Conference on*, pages 119 –122, sept. 2011.

[NSB11c]   M. Niknahad, O. Sander, and J. Becker. A study on fine granular fault tolerance methodologies for fpgas. In *Reconfigurable Communication-*

centric Systems-on-Chip (ReCoSoC), 2011 6th International Workshop on, pages 1 –5, june 2011.

[NSB12]   M. Niknahad, O. Sander, and J. Becker. Fine grain fault tolerance- a key to high reliability for fpgas in space. In *Aerospace IEEE Conference,* March 2012.

[NSC⁺11] M. Niknahad, O. Sander, L. Carro, J.R. Azambuja, J. Becker, and F.L. Kastensmidt. Using quadded logic in nanoplas to aggressively increase circuit yield. In *Nano, Information Technology and Reliability (NASNIT), 2011 15th North-East Asia Symposium on,* pages 180 –185, oct. 2011.